Praise for DV Filmmaking

"For university students of digital video production [...] others in its robust range and useful organization and illustration of topics. Aronson communicates his extensive expertise with extraordinary attention to detail, a wealth of familiar examples, and clarity that reflect his experience as a teacher, a lifelong student of film and video, and a professional filmmaker/videographer."

—*Francine Shaw, Director of the Educational Communication and Technology Masters program at NYU*

"Ian Aronson offers an extraordinarily well-informed and thorough discussion of every aspect of digital production. Anyone contemplating shooting a film on DV—whether narrative or documentary, feature or short—would do well to swallow this book whole."

—*Robert Edwards, Writer and Director of* Land of the Blind

"This introduction to the world of digital media production offers a clear and comprehensive explanation of technical and aesthetic information critical to good moviemaking. It is a terrific bible for beginning makers and a useful resource for the professional."

—*Kristine Samuelson, Professor, Documentary Film and Video Graduate Program at Stanford University*

DV Filmmaking:
From Start to Finish

Ian David Aronson

O'REILLY®

BEIJING • CAMBRIDGE • FARNHAM • KÖLN • PARIS • SEBASTOPOL • TAIPEI • TOKYO

DV Filmmaking: From Start to Finish

by Ian David Aronson

Published by O'Reilly Media, Inc., 1005 Gravenstein Highway North, Sebastopol, CA 95472.

O'Reilly books may be purchased for educational, business, or sales promotional use. Online editions are also available for most titles (*safari.oreilly.com*). For more information, contact our corporate/institutional sales department: 800-998-9938 or *corporate@oreilly.com*.

Print History:

December 2005: First Edition.

Editor:	Steve Weiss
Production Editor:	Darren Kelly
Cover Designer:	Michael Kohnke
Interior Designers:	David Futato

 This book uses RepKover™, a durable and flexible lay-flat binding.

0-596-00848-1
[L]

This book is dedicated to:

The memory of my grandparents, David and Pearl Aronson.

Everyone who wakes up in the morning and pursues their artistic dreams.
More power to you.

Contents

Foreword

"If you don't know where you are going, all roads will get you there."

—Anonymous

The world of postproduction is a perfect vantage point for observing production and filmmaking. Postproduction—editing, sound, color correction, and output—is often nothing more than solving one problem after another to get a film finished and out the door. Many times one wonders how many of these problems could have been avoided, and time and money saved, had the producers or directors paid some attention to a few basics.

Possibly the biggest hurdle in any filmmaker's path, it seems to me, is often his view of his own creativity. I have never met a filmmaker who has said, "I have a so-so idea for a movie." All filmmakers have always told me that they had a really great idea for a film they were working on. I am sure they are being honest. However, this sense can often blind most producers and directors to a need to at least understand the tools and use them correctly to tell a clear story.

There's a flip side to this as well. Producers and writers who know their filmmaking tools well, and yet lack any sense of drama and creative ideas.

Some years ago, Mario Puzo, author of *The Godfather*, wrote a book called *The Godfather Papers* in which he described his process of writing, and in which he recalled his experiences in the film business. Puzo related an anecdote to point out the illusions that often surround creativity and writing.

Puzo was a private in the Italian army, and at the end of WWII, he was stationed in Germany, where he met up with a group of Russian soldiers. This particular Russian regiment was from a far-off Asiatic province, and the soldiers had never seen plumbing. They were amazed to see water flow out of faucets. One of these fur-hatted Russians ripped a faucet off and nailed it to a piece of wood outside his tent. When he woke up the morning and opened the tap, nothing came out. The Russian soldier was very disappointed.

Puzo's point: Some people think words flow out of a pen.

Flash forward a little. In 1984, the Mac arrived, courtesy of Apple Computer, and Adobe soon came out with PostScript, a technology that allowed for superior graphics and high-quality text printing on home laser printers.

WYSIWYG (What You See Is What You Get) was a completely new concept and created something called "desktop publishing." Predictions were made that desktop publishing would make a novelist out of everyone. People would sit around, churning out novels from their laser printers. Cultural pundits predicted the demise of publishing empires and heralded the arrival of a new age of authors. Alas, the dream turned out to be a vapor. All that came out of most people's laser printers were bake-sale posters with awful clip art and mismatching fonts.

Conclusion: Talent doesn't flow out of computers.

Let's go to the year 2005. The digital video revolution is in full swing. "Desktop editing" is here. Hey, anyone can make a great film!

Oops.

You get the point, I think. The old adage "the more things change, the more they remain the same" still holds true.

There's something sorely lacking in current books about filmmaking. Either these books talk about the tools, or the books explore the creative side of filmmaking. Here's one book that combines the two.

Ian David Aronson's book merges these two critical needs:

1. To understand the tools you are using.

2. To use other great creative ideas to draw from and understand how the process of creative filmmaking works.

By using examples from relevant and well-made films, Aronson effectively illustrates the use of the tools he discusses.

It may seem like a no-brainer to read "Use Headphones" during the book's tips on recording sound, but you'd be amazed at how often I have been challenged by materials with bad audio where the recordist simply put out a microphone without any facility (like a pair of headsets) to monitor the sound. Oversights like that are expensive to fix in post, and create unnecessary chaos.

At the same time, the creative side is not ignored in this book. Using examples of films that are either part of the cultural canon or are easily available, all of the techniques are well illustrated. I cannot imagine a better starting point for any filmmaker than to view the films Aronson uses as examples, while at the same time working to understand the tools described in the book.

Yes, you too can create good works. But first understand the tools well, and take time to learn from people who have already walked this road before you.

Happy filmmaking.

—Zed Saeed
Senior Post Production Consultant
DigitalFilm Tree
Los Angeles

Preface

Why Read This Book?

Making a good film is more affordable, and open to a wider range of people, than ever before. The reason is digital video.

If you have access to a digital video camera and a computer (you don't need to own them, you just need access), you can make professional-quality work. You don't need lots of cash, nor do you need a relative who owns a major studio. You just need a good story, and enough technical skill to realize your vision. This book walks you through all the details of making a video project, from start to finish, and shows you step-by-step how to master the advanced digital video technology at your disposal.

Audience

If you're inspired to make a film and share it with an audience, this book is for you. I walk you through the process of making a film: from lighting and framing a shot to setting up your own postproduction studio and refining your work into a finished product that you share with the world. Along the way, I describe scenes from films ranging from *Blade Runner* and *When Harry Met Sally* to *The Conversation* and *La Jetée* to introduce key concepts and explore examples of particular techniques. (I even reference the work of *Looney Tunes* creator Chuck Jones, because he used techniques that are highly relevant to digital postproduction.)

If you're a professional looking for a leg up, an educator who teaches digital media, or a video enthusiast looking to bulk up on your skills, this book was written with you in mind.

What This Book Is, and Is Not

First, and perhaps most important, this is not a book about a specific software program. This book is about using all the digital tools available to you to make the film you've always dreamed of.

The first half of the book explores the use of camera equipment, lights, and audio recording devices (and offers suggestions on how to rent them without going broke). The second half of the book discusses both Mac- and PC-based postproduction solutions. It also contains step-by-step exercises that walk you through advanced compositing, color correction, and audio finishing techniques using Final Cut Pro and After Effects.

Second, this book does not assume all readers share the same level of technical expertise or prior knowledge. This text is designed to be thorough, and to accommodate a variety of learners and learning styles. Some parts of the book focus more on film theory and conceptual knowledge, others are more technical. This course is designed for readers with extensive backgrounds in film production, as well as people new to the field and eager to learn.

What This Book Covers

As mentioned above, this book is divided into two parts. The first part covers both preproduction (the phase where you're planning your project) and production (the phase where you're actually out in the world using a camera). The second part covers postproduction: when you return to the studio (or your living room) to edit your footage and create a finished project. The second part of the book also explores distribution possibilities and ways you can get your film shown to an audience.

This book comes with a DVD that contains original, professionally shot video clips that you can use to complete the exercises in this book, and to practice the editing and effects creation techniques the book covers. The DVD also includes After Effects and Final Cut Pro project files containing completed versions of each exercise. In addition, the DVD contains a short film, titled *Big Luca*, that was produced especially for this book and employs all the techniques covered in this text. Right now, you're holding everything you need to learn the art of digital video.

Let's take it from the top.

Conventions Used in This Book

The following typographical conventions are used in this book:

Plain text

> Indicates menu titles, menu options, menu buttons, and keyboard shortcuts.

Italic

> Indicates new terms, film titles, URLs, email addresses, filenames, file extensions, pathnames, and directories.

We'd Like to Hear from You

Please address comments and questions concerning this book to the publisher:

O'Reilly Media, Inc.
1005 Gravenstein Highway North
Sebastopol, CA 95472
(800) 998-9938 (in the United States or Canada)
(707) 829-0515 (international or local)
(707) 829-0104 (fax)

We have a web page for this book, where we list errata, examples, and any additional information. You can access this page at:

http://www.oreilly.com/catalog/dvfilmmaking

To comment or ask technical questions about this book, send email to:

bookquestions@oreilly.com

For more information about our books, conferences, Resource Centers, and the O'Reilly Network, see our web site at:

http://www.oreilly.com

The author of this book also maintains a weblog at *http://weblogs. oreilly.com/* where he offers ongoing perspectives on digital technology in the world of film.

Acknowledgments

This book is the result of considerable effort by many people, and I owe a debt of gratitude to each one.

In particular, I would like to thank Executive Editor Steve Weiss, for going to great lengths to help me bring this book from an idea to a reality, and publisher Tim O'Reilly, for offering me the opportunity to work with O'Reilly Media. A number of other people made extraordinarily valuable contributions to this project, including copyeditor Linley Dolby, technical editor Bonnie Blake, editorial assistant Michele Filshie, senior technical illustrator Robert Romano, Zed Saeed who wrote the exceptional forward, and the many other people who worked with me to make this book worth reading. Thank you all very much and I look forward to working with you again soon.

I'd also like to thank my agent, Margot Maley Hutchison at Waterside Productions. I have no doubt that this book would not exist without your help.

In addition, I'd like to thank Jan L. Plass and Francine Shuchat Shaw for continually going out of their way to help me in my post-graduate work at

NYU, and I'd also like to thank everyone else who has helped me at one point or another in my career. There are far too many to list individually, but I would be remiss if I didn't acknowledge the work and energy of Michael W. Dean.

Most of all I would like to thank all my friends who have been such a big part of my life for so long, and a few new ones (especially *ma cherie* So-Young, *la co-star de mon film de Disney*).

IAN DAVID ARONSON
NEW YORK CITY, 2005

The Freedom of Digital Video

Film is expensive. Shooting a feature-length—or even a short—film means spending money to purchase film stock, paying to process the film once it's been exposed, and then spending more money to create and distribute prints after a film has been edited.

Video, on the other hand, is both flexible and affordable. You can shoot, edit, and output your project for a fraction of the costs of working with film. Unfortunately, for many years, analog video was just plain ugly. While film offered remarkable depth and clarity of color, video images were flat, had harsh edges, and colors that didn't look natural. Film was the only medium suitable for people aspiring to cinematic excellence. Even audiences with no technical background could differentiate between video and film, because video simply didn't look as good. When *Happy Days* stopped shooting film and started using videotape, lots of people in the audience saw that something had changed, even if they couldn't identify what it was. To many viewers, video equaled cheap production and only film signified quality.

Until recently, the most selective film festivals, awards competitions, and theatrical distributors considered only work shot on film. Because of the expense, however, film production remained beyond the reach of many directors. As a result, many good ideas never made it to the screen, and many great video projects never made it to an audience. Independent producers pined for a medium that would let them shoot film-quality images for the cost of shooting video.

Enter the digital camera. In the mid-to-late 1990s, filmmakers discovered affordable mini DV cameras that captured surprisingly high-quality images. These cameras, such as the Sony VX1000 and Canon XL-1 (Figure 1-1), fell into a relatively new category of products known as *prosumer*—they blurred the line between high-end consumer equipment, and lower-end professional gear. Prosumer equipment employed newly available digital technology to produce $3,000 cameras that rivaled the quality of professional analog video cameras that cost 10 times as much. The digital camera, followed shortly by affordable digital editing systems, created a new age of opportunity for the independent filmmaker. People could now realize their film ideas on a low budget without sacrificing quality. Today, independent producers use digital technology to create academy award–winning documentaries, theatrically successful dramatic films, and everything in between. As digital technology improves, prices drop, and new opportunities present themselves to independent filmmakers across the globe.

Figure 1-1. Affordable mini DV cameras like the Sony VX1000 and the Canon XL-1 enabled filmmakers, for the first time, to shoot professional-quality images at consumer-equipment prices.

SIDEBAR

Linear Versus Nonlinear Editing

From its conception, video was a linear medium. That is, to get to the end of a videotape, you had to fast-forward through the beginning and then the middle. Now, audiences watching a DVD can skip to their favorite romantic love scene or testosterone-packed action sequence at the touch of a button. Watching video no longer follows a linear progression, and neither does editing.

An editor working in Apple Final Cut Pro, or another desktop video application such as Avid Express DV or Adobe Premiere Pro, can navigate to any part of a project and add or trim footage without disturbing the edits that come before and after. (For more details, see Chapter 9.)

Another tremendous advantage of nonlinear editing systems is the effects they enable you to easily add to your work. In the recent past, an editor working in analog video needed a separate digital video effects unit (DVE) to create any type of effects, including even simple transitions. Now, nonlinear editing systems routinely ship with sophisticated effects capabilities—including the ability to create the simulated camera movements in Chapter 13 as well as the chroma key effects described in Chapter 5 and Chapter 12.

The Flexibility of Digital Recording Versus the Expense and Constraints of Film

Digital video is, without any doubt, a flexible medium. From the way you shoot to the way you edit and store your footage, digital video offers a tremendous number of options. Digital video is, in fact, so flexible that it's changed the way people make movies.

Length

Digital video tapes are widely available in lengths of 6 to well over 120 minutes. In terms of both price and sophistication, digital video equipment ranges from consumer categories to high-end professional uses. The most widely used format is mini DV, found in both consumer camcorders and prosumer equipment. The mini DV format is often abbreviated as simply DV, in part because it's so prevalent. Other digital video equipment and tape formats include DVCAM, DVCPRO, and Digital Betacam. These more sophisticated formats use differently formulated tapes and record increasing amounts of information, resulting in more detailed images, but you can still do great work using mini DV equipment.

Shooting with a longer tape in the camera enables a cinematographer to record for extended periods of time, often an entire hour or more, without stopping. This has created an entirely new style of shooting for documentary filmmakers who can turn the camera on and seamlessly capture *everything* that happens around them.

For *Metallica: Some Kind of Monster*, directors Joe Berlinger and Bruce Sinofsky spent two years filming members of the metal band, Metallica, in various endeavors, including 50 group therapy sessions that lasted four hours apiece. The filmmakers used two DVCAM cameras, a high-end Sony DSR-500, and a prosumer Sony PD-150 to record 1,600 hours of footage that reveals band members in some remarkably unguarded moments. Unlike producing a fiction film where the directors know the outcome in advance, "we work in the exact opposite way," Berlinger told *indiewire.com* (an excellent web site for news on independent film). The directors sifted through their footage in postproduction to find the most important scenes and then structured the film around those clips—a working style that is made possible, in many ways, by the introduction of digital video.

In *Super Size Me*, director Morgan Spurlock spends 30 days in front of a camera, eating nothing but food from McDonald's for breakfast, lunch, and dinner (he says he only supersized when people asked him to). Using a Sony PD-150 camcorder, Spurlock and his director of photography, Scott Ambrozy, traveled together as a two-man crew. He told *Filmmaker* magazine that working in digital video helped keep costs down and provided the flexibility they needed to capture his entire experience on camera: "We

Metallica and DV

On a recent summer afternoon, I went to my local art house multiplex, the Sunshine on East Houston Street, to see *Metallica: Some Kind of Monster*. While waiting in the lobby for the next showing, I learned that director Bruce Sinofsky was in the theater having an informal Q & A with the audience from the earlier show. I showed the usher my ticket and asked if I could go in to hear the director. He said no. When he wasn't looking, I walked in, found a seat, and asked Sinofsky how the project would have been different if he and Berlinger had tried to shoot film. (Before working on *Metallica: Some Kind of Monster*, Sinofsky and Berlinger achieved great success as documentary filmmakers with *Brother's Keeper* in 1992 and *Paradise Lost: The Child Murders at Robin Hills* released in 1996, both of which were shot on film.) Without hesitating, Sinofsky answered that had he not used digital video, his film "would not have been possible." He and Berlinger recorded everything that happened, and later, selected only the most powerful material to include in the final product. This would have been too cumbersome and too expensive to do with 16 mm film. On his way out, he gave me a promotional guitar pick with a graphic from Metallica's latest CD on one side and the logo for his film on the other.

filmed nonstop for six or seven weeks during my eating phase." Shooting constantly for an extended period of time enabled Spurlock to record every interaction and eating experience—even the unseemly ones—then decide which he wanted to share with an audience. Nothing had to be ignored.

Shooting film is an entirely different process. 16 mm film, the format traditionally used by independent filmmakers, comes in 400 foot reels, which only last about 10 minutes apiece (100 foot reels are also available, but these are even shorter). Because 16 mm film limited documentary filmmakers to shooting only 10 minutes or so at a time, they often missed a fair amount of action. To compensate, filmmakers would often conserve their film, turning the camera on only when they were certain something was worth shooting, and the rest of the time, they would record only audio. In postproduction, editors would then weave different shots together, along with the recorded audio, using a patchwork of material to cover gaps in the footage.

NOTE

Regardless of flexibility and cost, there are still filmmakers—and even a number of viewers—who believe video will never look as good as film. As a result, some directors continue to shoot film but edit using a nonlinear editing system. In Appendix A, I interview two successful filmmakers who, for various reasons, shoot film and use digital technology in postproduction.

In the film era, very few projects, documentary or narrative, contained continuous shots longer than a few minutes, simply because the technology made it so difficult. Orson Welles opens his 1958 film, *Touch of Evil*, with a single continuous shot of the characters winding their way through a Mexican border town. The shot is extraordinarily well choreographed, and extremely risky given the technology he was using—think of how expensive it would be to stop and redo everything if someone made a mistake several minutes into the shot (I discuss this film in more detail in Chapter 17). These days, thanks to DV, lengthy continuous shots are increasingly common, and if you run out of tape while you're shooting, you can just pop a new cassette into your camcorder. Because directors don't need to worry about continuously changing reels of film, they can let actors improvise for longer periods of time, record multiple camera angles, and shoot numerous takes of a scene to make sure everything works exactly the way the want. If a scene runs longer than anticipated, that's okay; tape is cheap. If an actor makes a mistake, retakes are no problem. To a director shooting film, it's a different story: if the camera runs out of film, the director's pretty much out of luck. Changing a reel of film is a difficult process, requiring complete darkness, exceptional care, and a skilled professional. If a crew member accidentally exposes the film to daylight or incorrectly threads the film through the camera, the work of everyone on set can be instantly ruined.

The cost and anxiety of film processing

Good quality 60-minute mini DV tapes retail for less than $10 apiece and cost half that when you buy in bulk. A 10-minute reel of film, in contrast, can easily cost $100. The expenses don't stop there—the real cost of shooting film lies in lab fees. Once you shoot film, you then need to process the negative, which costs you even more money before you can see your work. Processing 16 mm negative costs about 20 cents a foot, so a 10 minute reel of film costs about $80 to develop (400 feet of film at 20 cents per foot). Since you can't watch a negative to see what your footage looks like, you then need to create a *workprint*, which serves as your viewing copy (workprints are often referred to as *rushes* or *dailies*, because they're quickly processed film prints used only for initial screenings and editing, and they aren't shown to audiences). Workprints cost an additional 32 cents a foot, so before you can even bring that 10 minute reel of film into the editing room, you have to pay another $128 in processing (400 feet at 32 cents per foot) on top of everything you've already spent.

The scariest part of all this is you won't know if your film looks good—or is even usable—until after you've paid for each step. If you add up the previously listed costs, including the initial film purchase, it comes to more than $300 for each 10-minute reel. If you've shot an hour's worth of film, you can easily spend more than $1800 before you can see what you've got, and what you have may not be what you want. In fact, it may not even be usable. The image you see when you look through the viewfinder of a camera is often not the same as what shows up on film. You might not notice exposure problems, focus issues, or a stray hair caught in the gate (the part of the camera that film passes through as it's exposed) until after you've finished shooting and you get your film back from the lab.

Shooting DV is a much more straightforward process—what you see in a well-calibrated field monitor or viewfinder is exactly what's reordered on the tape. At the end of each take, the director of photography knows what he's recorded. If something isn't right, you can easily identify the problem and fix it in a retake. With film, you won't see the results of your work until the next day (at the earliest) when the negative is processed, so the same problem might present itself in reel after reel of film but go entirely unnoticed during shooting.

Video tolerates mistakes

As a medium, digital video is far more forgiving than film. Higher-end video equipment offers a range of technical options and requires you to make significant decisions that shape the outcome of your project, but if you make a mistake, chances are you can fix it in postproduction. Filmmakers often come back from a shoot and realize that something went wrong (*Murphy's Law*, that whatever can go wrong will, is a big part of life for the independent producer), but DV is flexible enough that you can fix a wide range of problems when you edit your footage. DV tape, like anything else, still has its limits. Repeatedly rewinding a cassette to watch a segment in your camcorder can physically damage the tape, ruining your footage. Because DV tapes use a magnetic process to record information, you can also erase a tape by accidentally passing it through a magnetic field (you may be amazed to find how many magnetic fields there are in your home; keep your tapes

Ross McElwee and the 16 mm Personal Documentary

The personal documentary, in which the director of a film is also the star, existed long before the advent of digital video, but it took a very different form. In 1986, director Ross McElwee shared the intimate details of his life with the audience in *Sherman's March,* which The New York Times called "disarmingly engaging." At the start of the film, McElwee says he intended to make a film retracing General Sherman's devastating march through the South during the Civil War, but the film quickly becomes a self-examination of McElwee's own love life. McElwee tells his story by pairing footage of his travels with warm and brilliantly funny personal narration. The on- and off-camera narration is both a key to the film's success and partly a device to work around the technical limitations of film production. (Spurlock employs narration to great effect in *Super Size Me,* but he's also working with significantly more footage, including infinitely more private moments than anything that appears in *Sherman's March.*) Because McElwee worked with reels of film, as opposed to 30- or 60-minute DV tapes, he simply didn't have the option of capturing everything that happened and later sorting through it in postproduction as filmmakers commonly do today. Instead, he employed a disciplined, carefully structured approach to his filming. Later, during editing, he wrote narration to hold his film together and segue from one sequence to another.

away from the TV). Additionally, DV cassettes are vulnerable to human curiosity—sometimes people can't help themselves, so they fold back the protective cover guarding the tape itself, leaving their valuable footage exposed to dust and other contaminants.

SIDEBAR

The Vulnerability of Any Recorded Media

A tape can fail, and if it does, you lose everything on it. Filmmakers have adopted a variety of approaches to compensate for this. Here are a few suggestions:

- Resist the temptation to watch your footage in the camera. Rewinding and fast-forwarding a DV tape in the camera causes excess wear and tear, which can lead to a wrinkle or even a rip in the tape. If the tape is damaged, there's no way to repair it.

- Capture your footage using a deck, instead of a camera; this can help you avoid damage to your tapes. Cameras are not built to safely rewind or fast-forward tapes, decks are.

- When you capture using a deck (or even if you capture directly from the camera if a deck is not in your budget), you lessen the chances of damaging your tapes if you capture your footage without repeatedly running through the tape to find the exact start or end of a shot—just capture a little extra and edit later.

- Once you've captured your footage, archive it to an external disk drive. This way if the original tape is ever damaged, you already have a saved version of your footage.

- Make a duplicate copy of each tape. Because you're working with digital information, you can create backup copies of your footage without any loss in quality (in fact, copies of digital tapes are called clones because they're identical). Once you have a backup copy, you can place the duplicate tape on a shelf and keep it in a safe place in case something happens to your original.

- If you know you'll never get the opportunity to record something again, you might consider trying a multi-camera shoot. A multi-camera shoot may complicate the logistics of your operation (see the note on one camera versus two camera shoots in Chapter 3), but you'll definitely walk out with more than one tape.

For more information on capturing and storing video, see Chapters 9 and 10. No matter how you slice it, film is far less forgiving, and technical problems can easily leave you with unusable footage that you spent tremendous time, energy, and money to capture. Because film requires processing before you can view it, it's also vulnerable to environmental hazards and lab mistakes while DV is not. When you take a DV tape out of your camera, it's ready to work with, and if you handle the tape carefully, you can more or less be assured your footage is safe. Once film is exposed in the camera, it's highly vulnerable until it's processed. A friend of mine recently traveled across country to shoot the latest installment of an ongoing documentary project and returned with 10 reels of film. Shortly after dropping his film off at the lab he got a call that, due to a problem with the lab's equipment, several reels of his film had been damaged, and the last two had been rendered unusable.

These were, of course, the reels containing footage that was the hardest to get, and as a result, the hardest to reproduce. This doesn't mean digital video is immune to mistakes (it isn't—see the sidebar, "The Vulnerability of Any Recorded Media") or that 16 mm is an antiquated magnet for technical errors, but DV clearly is a more cost effective, straightforward production medium, and it's changed the world of independent film forever.

Digital Features: A Brief History of Directors Who Chose Digital Production over Film

When digital video cameras first became available, people who were already shooting analog video flocked to the new technology, but film-oriented directors remained hesitant. Years ago, I went with some friends to see *The Cruise*, a feature that garnered significant attention on the festival circuit as one of the first projects shot and edited using entirely digital equipment. Everyone I went with liked the movie, but one friend (who to this day is still a particular devotee of 16 mm film) looked at us afterward and said, "You know, it's still video."

Modulations: DV interspersed with film

Even in an all-digital context, many filmmakers doubted the quality of video in any form. Cinematographers earn a living by making the world look good on screen, and many were reluctant to trust a DV camera if it meant risking their reputation on an unproven technology that might not produce the greatest possible image. Film is, above all else, a visual medium.

One of the first filmmakers to use DV in a large-scale documentary was Iara Lee, who used DV equipment to shoot interviews for her 1998 film *Modulations*, which explored the world of electronic music. To produce *Modulations,* Lee shot interviews with musicians, producers, and DJs around the globe, then combined the edited interviews with observational footage shot in clubs, at concerts, and at other venues. Lee is highly conscious of the importance of quality images, and had shot her previous work *Synthetic Pleasures* entirely on film. In *Modulations*, she used digital video to shoot her interviews and shot her observational sequences entirely on 16 mm. Lee's films are strikingly visual in their approach, and they're fun to watch because of her innovative use of styled imagery.

Shooting DV interviews allowed Lee to save considerable sums of money. On-camera interviews require a director to shoot far more footage than what winds up on screen in the finished product. Often, only minutes of an hour-long interview make it into the edited version, and many interviews get nixed altogether. Lee filmed hundreds of interviews for *Modulations* and told *Res* magazine (*http://www.res.com*, which has been covering digital video since its inception) she would not have been able to afford the cost of shooting them on film. Placing the interviews alongside 16 mm observational material that she edited and manipulated on an Avid digital editing system helped her achieve the look and style she wanted without going beyond her budget.

Buena Vista Social Club: DV in place of film

In this project, released theatrically in 1999, Director Wim Wenders used a combination of prosumer, consumer, and high-end professional digital video cameras to shoot in environments where traditional 35 mm or 16 mm film would not have worked. *Buena Vista Social Club* documents a Grammy-winning group of Cuban musicians as they record a new album and travel to New York City to perform at Carnegie Hall. The entire film is beautifully shot in digital video, including a number of sequences in a recording studio, where the mechanical sound of film running through a camera would have interrupted the work of the musicians who appear on screen.

The bulk of the film was shot in Digital Betacam, which was, at the time, the top quality digital format available. Wenders employed a cinematographer for the principle camera work and recorded additional footage himself using a prosumer DV camera, as well as a very small consumer DV camera. "This way I could sometimes shoot in places and situations where you'd just never get with a film camera, even a 16 mm," Wenders is quoted as saying on his web site, *www.wim-wenders.com*. "This film could have never been made, such as it is, on film. This is truly a product of the new possibilities we have as filmmakers with the digital tools."

The power of *Buena Vista Social Club*, in addition to the music, is the film's warmth and its ability to make you feel close to the musicians it portrays. I've shown this film to several classes as an example of filmmaking that reaches audiences on an emotional level. Much of this is due to Wenders's ability as a director—his filmmaking doesn't get in the way of his story—but the film is also aided by the technology Wenders used in production. The flexibility of digital filmmaking enables directors to work with very small crews and create an intimacy that audiences can see in the final product. Digital cameras don't require crews to set up cumbersome or intrusive lights, and they can be used for such extended periods of time that people tend to relax and almost forget a camera is in the room.

Southern Comfort: digital immediacy

Southern Comfort, Kate Davis's film about a female-to-male transsexual fighting a losing battle with ovarian cancer, won the Grand Jury Prize for

All About the Ratio

A shooting ratio is how much footage you shoot in relation to the length of your finished product. For example, if you shoot six hours of tape to make a one hour movie, you have a six to one ratio, which is written 6:1. Depending on budget constraints and shooting style, filmmakers commonly use ratios ranging from 4:1 up to 12:1 or more.

Shooting more footage gives you more material to work with, but it also increases your costs (both in terms of raw stock and the time you spend sifting through your tape in postproduction). Ultimately, any shoot is a balance between what you want and what's available given your circumstances. Bay Area filmmaker Les Blank (director of *Garlic Is As Good As Ten Mothers* and *In Heaven There Is No Beer?*) is often quoted as saying that shooting is not aesthetic, it's athletic.

I was on a shoot one day with a friend who said, "You know, it's all about the ratio." His point was that every shot you take doesn't need to be magic, you just need to shoot enough so you can cut out everything that isn't.

Documentary at the 2001 Sundance Film Festival. Davis, a professional film editor with extensive experience in 16 mm film, shot the project herself in DV. She used a small Sony VX1000 camera that looks remarkably like a standard home movie camcorder. Davis often worked on this film alone, and because she used a small camera that didn't draw attention to itself, people felt very comfortable having her around. Amy Taubin wrote in the *Village Voice* that the "unobtrusiveness" of Davis and her camera "proved invaluable." Davis "is treated by everyone on the screen like a dear friend," according to Taubin. "They talk to her as if they were unaware of the camera and simply want to include her in the conversation." Creating an environment where people feel comfortable in front of a camera is important for any film, but especially important to a film about a controversial subject. Working in DV helped Davis shift the focus away from herself and develop a close relationship between the audience and the film's protagonist, Robert Eads.

The lower cost of digital production also allowed Davis to go out and start shooting instead of endlessly having to fundraise to cover the expenses of working with film. "Ms. Davis had brains enough not to wait for funds because she knew she had a great story right in front of her," Elvis Mitchell wrote in the *New York Times*. Since *Southern Comfort* chronicles the last year of Eads's life, it's quite possible that if Davis spent her time fundraising instead of filming, the project might never have been completed.

The Hotel Upstairs: digital accessibility

Digital video has not only created opportunities for recognized filmmakers like Wenders and Davis, but has opened doors for filmmakers at the start of their careers. Traditionally, newly minted film school graduates went to work for established filmmakers and, using their bosses' equipment after hours and on weekends, slowly began to create works of their own. Today, film school graduates still work for other people to earn money, but DV technology has enabled many to start making their own projects right away, without the oversight of a more senior filmmaker. (The same technology has also enabled people to make great films without ever attending film school or working for another filmmaker—it's created a whole new world of access.)

The Hotel Upstairs, a portrait of a San Francisco residential hotel and its long-term occupants, was made by director Daniel Baer shortly after he graduated film school, using DV equipment he purchased with a small grant and an editing system he bought with friends. Baer and some former classmates jointly purchased a used Avid digital editing system. Before the introduction of high-quality software-based editing systems like Final Cut Pro, nonlinear digital editing required the use of expensive hardware-based systems, like the Avid, which used specially modified computers in addition to software and generally cost more than $60,000. Each time Avid came out with newer models, the value of the older models dropped to a small percentage of their original price, because the hardware became permanently outdated. Baer and three friends purchased an older Avid with money they scraped together, and he used the system to edit his project. (As I was writing this section, a friend emailed me that the Echo Park Film Center in Los Angeles was selling an old Avid for $125. Baer paid significantly more for his Avid back in 1998, but it still cost much less than the original retail price.)

Baer had directed two successful 16 mm student films before starting *The Hotel Upstairs*, but shot the project in digital video because he could afford to, and because DV performs better in low-light situations than film. The rooms of the hotel were too small for Baer to set up lighting equipment, but even if they had been big enough, he feared lights would have created an unnatural situation. "I wanted to shoot without lights to give subjects the idea there wasn't a hard line between their life, and their life when I was shooting."

Baer purchased a newly available Canon XL-1 digital camera and used it to shoot the film himself. The XL-1 retailed for slightly more than $3,000 but recorded images comparable to cameras that cost significantly more. Baer used additional money from the grant to cover his share of the Avid. To affordably give his work a more "film-like visual quality," he shot exterior views of the hotel in 16 mm, and incorporated them into his project along with sequences of still images shot with a 35 mm still camera.

Traditionally, independent filmmakers paid rental fees by the day or week for cameras and editing equipment that they couldn't possibly afford to purchase, but because Baer now owned a high-quality camera and a professional editing system, he was able to create a documentary without spending a lot of money. *The Hotel Upstairs* premiered at the 2001 Los Angeles Film Festival and went on to screen at festivals around the world.

Star Wars Episode II: Attack of the Clones and *Call AT&T*— big budget digital

At the other end of the scale, big-budget Hollywood directors like George Lucas, and high-end advertising agencies such as Young and Rubicam, have embraced digital video. Lucas shot *Star Wars Episode II* using *High Definition* (HD) digital video cameras specifically developed for the latest installment of his sci-fi epic. Young and Rubicam, the advertising agency for AT&T, used similar HD technology two years later to film a series of commercials staring comedian Carrot Top.

High Definition is currently the highest quality digital format available, and many filmmakers are attracted to HD because it uses the same frame rate and aspect ratio as film, which helps make it look more like film at a much lower per-minute price. (Frame rate and aspect ratio are described in detail in Chapter 2.) Because of the technical similarities between HD and film, material shot on HD provides directors with a more cinematic look than other forms of digital video and a more efficient transfer to 35 mm film for

NOTE

At this point, the main price difference between mini DV and HD is the equipment you need to shoot and capture. While applications like Final Cut Pro and Premiere both offer affordable HD editing capabilities, HD cameras and decks cost significantly more than their mini DV counterparts. At the time of this writing, a Panasonic HD deck compatible with Final Cut Pro retails for more than $20,000 and a compatible Panasonic HD camera retails for more than $65,000. Prices will likely come down in the near future, but for the moment, high quality HD is still outside the purchase price range of most independents.

Newer versions of Final Cut Pro are compatible with a format called HDV, which records compressed high definition video onto a mini DV tape. HDV cameras generally fall into the prosumer price range. For example, the Sony HDR-FX1 retails for less than $4,000.

projection in theaters. *Star Wars Episode II* was projected digitally in theaters equipped with digital projection systems, and on 35 mm film in those without. AT&T created digital versions of its "Call AT&T" commercials for television broadcast and 35 mm film versions to screen in theaters between coming attractions.

Lucas, who has become known for computer-generated environments, characters, and action sequences in his recent work, chose HD because it provided him with an unprecedented level of control. "It's a much more malleable medium than film, by far," he told *American Cinematographer* magazine. "You can make it do whatever you want it to do, and you can design the technology to do whatever you want to do." (You can view selected *American Cinematographer* articles for free online, at *http://www. theasc.com/magazine/.*)

HD is generally a more expensive format than other forms of digital video but still costs far less than shooting in film. In addition to increased affordability and control, like other digital formats, HD also lets directors shoot longer takes and immediately see the results of their work. "HD won't bite you," Ken Yagoda told *Millimeter* magazine. Yagoda, Young and Rubicam's Managing Partner/Director of Broadcast Productions, said HD saved time not only because production is very efficient, but because directors didn't need to guess about what they did or didn't get on film. "One of the advantages is that you really do see what you're getting, so you can safely wrap a sequence and move on with a great deal of confidence."

> **NOTE**
>
> *Millimeter is a very cool and easy-to-read magazine about post-production techniques and effects creation. You can read it online at www.millimeter.com, or subscribe to the print version of the magazine at no charge in exchange for submitting your demographic information (see millimeter.com/ subscribe for details).*

November: big screen HD success with a prosumer DV camera

Unlike *Star Wars Episode II*, which was shot with custom-designed HD cameras far beyond the price range of an independent, *November*, a dramatic film staring Courtney Cox, was shot on a mini DV camera that retails for less than $3,000. *November*, which won the 2004 Sundance Excellence in Cinematography award, is an example of a growing number of successful art house films shot in a prosumer format and then released on film in theaters. The film's director of photography, Nancy Schreiber, used a Panasonic AG-DVX100 camera, carefully lit each scene, and manipulated the camera's exposure and focus settings to create a visual feel that Sundance judges were obviously pleased with, and *Filmmaker* magazine says, "changed the look of mini DV."

The AG-DVX100 uses the same mini DV format as the affordable cameras used by Davis and Baer, but is one of few prosumer cameras that can be set to record at the same frame rate as film. Using this frame rate allowed *November*'s producers to smoothly transfer the finished product to 35 mm film for theatrical distribution. The HD cameras Lucas used to film *Star Wars Episode II* also recorded at the same frame rate as film, but did so at a much higher cost. The shooting budget for *November* was only $150,000, according to *Filmmaker*, and the budget for the entire film was only $300,000. This may seem like a substantial sum of money if you're funding a production on your own. In the world of theatrically released features, however, films that cost $5 million are often described as low budget.

November's production company InDigEnt, which stands for independent digital entertainment, also produced the theatrically released film *Tadpole* staring Sigourney Weaver and John Ritter, which was shot in digital video and edited in Final Cut Pro. *Tadpole* was such a successful film reviewers didn't even mention it had been shot in DV—the *Hollywood Reporter* called it a "scrumptious amusement." Technological advances in digital filmmaking now allow directors to create low-cost, high-quality projects that people respond so well to, audiences don't even realize they're watching video. (I knew this was a film I had to see when I went to see another film and heard *Tadpole*'s audience laughing from outside the theater.)

Film is still expensive, but the ugly-duckling format of video has grown into a beautiful swan.

Let's put it to work.

This book is structured to provide an overview of digital video technology in the opening chapters, and then move into more specific details and techniques later on. This chapter explored the benefits of working in digital video (it's both affordable and flexible) and also provided a brief history of directors who chose to work in video over film, and why. Chapter 2 explores technical fundamentals such as *aspect ratio*, *frame rate*, and the difference between *lines of resolution* in a video monitor and *pixels* in a computer monitor. Understanding each element is an essential step toward producing a good video project, and the earlier you get a handle on them, the earlier you can start making the film of your dreams.

Digital Cinematography

2

Working in film and video is by definition a technical process: a filmmaker uses the technology of cinema to realize her ideas and share them with an audience.

As a filmmaker using digital video, you have a significant range of technical options. While these choices and details may seem overwhelming at first, they're important to think about because each decision ultimately shapes the outcome of your film. The choices you make while planning and shooting a film determine the options available in postproduction, and choices in postproduction either limit or expand your distribution options. A solid understanding of digital filmmaking's nuts and bolts can go a long way toward helping you achieve your creative vision.

The first step is defining the technical standards for your project. This section provides an overview of aspect ratio, frame rates, and video standards, and addresses the impact each choice has on the final version of your production.

Aspect Ratio

It's no secret that a television screen is a different shape than the screen in a movie theater. Both are rectangles, but the traditional shape of a television is closer to square, and a movie screen is significantly wider than it is high. The technical term for the difference in shape is *aspect ratio*, which means the proportional relationship between width and height. Traditional forms of video use a 4×3 aspect ratio, and widescreen video uses 16×9. Figure 2-1 shows examples of both.

Figure 2-1. The photo at left appears as a 4x3 image, and the photo at right is shown at 16x9. Even though they depict shops on the same street corner, the difference in aspect ratio produces markedly different compositions.

Aspect Ratios, in Detail

While 16x9 is commonly referred to as widescreen there are, in fact, a number of other widescreen aspect ratios. Each of the following wide-screen aspect ratios has an even greater width to height ratio than 16x9. If you shoot at any of these aspect ratios, you'll need to make some adjustments to fit your work on a 4x3 or even a 16x9 screen.

- 35 mm film displays at a ratio of 1.85x1, which is slightly wider than 16x9.

- Widescreen Anamorphic, which has an aspect ratio of 2.35x1, requires a special lens on the camera and another special lens on the projector. This extremely wide aspect ratio creates a frame that's more than twice as wide as it is high (approximately two and one third units of width for every one unit of height).

- 70 mm film, used to shoot widescreen epics such as *The Poseidon Adventure, Apocalypse Now,* and *The Thin Red Line,* displays at 2.0x1.

A 4x3 aspect ratio means that for every four inches of width in a screen, there's a corresponding three inches of height. For example, a television screen that's 20 inches wide would be 15 inches high. Screens conform to standard aspect ratios so that images maintain their proportions on any television set. Regardless of whether you watch your favorite TV show on a large or a small television screen, the width and the height of the images scale together in proportion. If they didn't, the effect would be like looking in a funhouse mirror: images would be stretched out of shape and distorted, so Fat Albert might look fat on some screens, but appear quite fit on others. Because programming is created and displayed at uniform aspect ratios, images are proportionally resized from screen to screen. Theatrically released films and widescreen video formats use aspect ratios that are very different than the television screen. These formats are almost twice as wide as they are high, which enables filmmakers to capture a much larger area of action in a single frame. Widescreen formats allow a director of photography to compose striking panoramic landscape shots and big-screen action sequences—for example shots of cowboys riding across the screen through the Badlands.

Difficulties arise when you have a project created in one aspect ratio but you want to exhibit in another, as shown in Figure 2-2. For example, if you shoot in a widescreen aspect ratio, you have to make some choices to fit your work on a 4x3 video screen. This has traditionally been a problem for directors who shoot 35 mm film and want to broadcast their work on television, or distribute via home video. One choice is to conform a widescreen shot to 4x3 by pushing the edges of the frame in toward the center, which allows directors to use the entire frame but makes everything slightly taller and thinner. This is often done to adapt the widescreen opening sequences of films for television broadcast. When this technique is used to convert extra widescreen film formats, such as 70 mm, it results in impossibly slim actors and titles that are almost unreadable because they've been distorted to accommodate a narrower screen.

Another method, called *pan & scan*, keeps a widescreen frame at its original aspect ratio but displays only a selected area wide enough to fill a 4x3 screen. This doesn't change the height or width of the images, it just cuts off material at the sides—anything outside the selected 4x3 area is eliminated. This does a disservice to the director of photography who composed the shot, because he thought the material at the sides of the frame was important enough to use. Another obvious problem with this method is deciding which part of the screen to keep and which to get rid of. It's called pan & scan because sometimes a technician will add an artificial pan that was

Figure 2-2. This illustration shows a widescreen image, from left to right, compressed to fit a 4x3 screen, pan & scanned so the edges are cropped, and letterboxed leaving blank space at the top and bottom of the frame.

not in the original shot to include story-critical action in the far left or far right of a particular scene. (And often, the film's director and director of photography are not consulted in the process.) A good film uses the entire frame, so the important action doesn't always show up in the same spot. Depending on the shot, you need to see different parts of the frame to follow the action. You've probably seen pan & scan used in movies adapted for television and may have found it a much less satisfying experience than viewing the original widescreen format.

A third solution is to *letterbox* the shot, which preserves the aspect ratio by proportionally shrinking an entire widescreen image to fit the width of a 4×3 screen, leaving strips of black at the top and the bottom. This is generally the preferred choice of filmmakers, because it preserves the integrity of each shot, but some viewers dislike any blank space on their television screen. (Letterboxing tends to be favored by cinema fans, people who love movies and know a lot about them, as opposed to casual viewers.)

Figure 2-3. The illustration shows a 4x3 image reformatted to fit a widescreen 16x9 aspect ratio. In the figure at left, the image is stretched horizontally to fit the wider screen space. In the figure to the right, the image appears in the center of the screen unchanged proportionally, with curtains of blank space on the sides.

Showing 4×3 content on a 16×9 screen is similarly difficult (Figure 2-3) —this has become a particular disappointment for people who purchase widescreen televisions and then notice that many television programs are still broadcast in a 4×3 format. To display 4×3 programming on a widescreen you could stretch your content, which would make the people in your show shorter and fatter, or you can leave empty space at the sides of your screen, called *curtains*. The simplest solution, as you may have guessed, is to shoot in the same aspect ratio as your primary distribution format. If your target audience is people watching television in their homes, 4×3 is a natural choice. If you want to exhibit your work in a theater, shooting 16×9 might be a better option.

Widescreen and DV

Anamorphic images have been used for decades in Hollywood movies. To create them, filmmakers attach a special anamorphic lens to a film camera. The lens distorts images to appear at an especially wide aspect ratio when they're played back on a projector that has been fitted with a similar anamorphic lens. If you have a video camera that accepts interchangeable lenses, you can use a similar process to capture anamorphic video (though you have to adjust for this in the editing; Final Cut and other high-end programs are capable of this).

Digital video cameras use a chip (technically referred to as a charged coupling device or CCD) to record images. Most mini DV cameras use one or more chips that record at 4x3. While these cameras can be set to record images in a widescreen format, they are incapable of recording true 16x9 media.

Some more advanced mini DV cameras contain an actual 16x9 CCD. These cameras can record an image at 16x9 and then save the image anamorphically on a mini DV tape. These cameras do not crop part of the frame, nor do they require an anamorphic lens.

If you're looking for additional information on aspect ratio and lenses, take a look at *http://www.centuryoptics.com/owning/faq.*

Anamorphic Video

The most widely available and most affordable format of digital video is mini DV, which uses a native aspect ratio of 4×3. This doesn't mean, however, that all video shot on mini DV will only work on a 4×3 screen. Many people, including Nancy Schreiber who shot *November*, use higher-end mini DV camcorders that offer a true 16×9 recording option to affordably shoot widescreen digital video. 16×9 video recorded on one of these mini DV camcorders uses a process called anamorphic video. To record 16×9 on a mini DV tape, a properly equipped camera records video using a 16×9 chip, and then slightly distorts the aspect ratio to fit the widescreen video into the mini DV aspect ratio. When anamorphic video is played back on a compatible monitor or editing system, it expands to its intended dimensions and displays as proportionally correct 16×9. A number of prosumer cameras have a good quality 16×9 setting, which enables a filmmaker to record 16×9 as described earlier. These include the Panasonic AG-DVX100A (the newer version of the camera Schreiber used), the Sony DCR-VX2100 (the updated version of the camera Kate Davis used to shoot *Southern Comfort*), and the Canon XL-2 (the contemporary version of the camera Daniel Baer used for *The Hotel Upstairs*).

Many less-advanced mini DV cameras offer a lower-quality 16×9 mode, which simulates widescreen footage by simply cutting off material from the top and the bottom of a 4×3 frame to create what looks like a letterboxed image. The result of these faux widescreen settings is lower-resolution video—it's just a 4×3 frame without the top and the bottom portions of the image.

To test whether your mini DV camcorder records anamorphic media, set it to 16×9 mode and look through the viewfinder while you're recording. If the images take up the full frame of your viewfinder and appear slightly stretched, your camcorder is recording anamorphic media. If the image appears with blank space at the top and the bottom of the frame, true 16×9 recording might not be an option for your camera (check the owner's manual for more details).

If, for one reason or another, you shoot at 4×3 and your finished project becomes so successful you need a 35 mm print for theatrical distribution, you have to use one of the techniques described earlier in this chapter to make your material fit a widescreen aspect ratio. If you know all along, however, that you plan to exhibit your work in a widescreen format (16×9 digital video or 35 mm film), you can simplify things by shooting 16×9 from the start. That way you'll never have to make a round peg fit in a square hole.

Frame Rate and Video Standards

Another item to keep in mind is *frame rate*: the number of images that appear on screen each second. Film and video simulate motion by displaying a series of still images, one after another. Research has shown that the human brain can process a limited number of images at a time, and that if a person views more than 15 images per second, the brain interprets the series of images as a single, moving image. This phenomenon is called *persistence of vision*. Without this, movies wouldn't work. They wouldn't seem lifelike to us.

Frame rates are measured in frames per second, or *fps*. Higher frame rates create smoother, more natural looking motion. 15 fps video looks jerky or choppy compared to 30 fps video, because a higher frame rate displays a greater number of images in the same amount of time, and the cumulative image sequence creates smoother looking motion.

Just as screens conform to standard aspect ratios for consistency, frame rates are standardized as well. If film or video is shot at one frame rate and played back at another, the motion doesn't appear natural, it looks sped up or slowed down. For example, a 15 fps shot of a person walking across the frame would go by twice as quickly if played back at 30 fps, so it might even look like the person is running.

To avoid problems with frame rate, aspect ratio, and other issues, video equipment conforms to specific standards. The broadcast video standard for North America is called NTSC and uses approximately 30 fps (29.97 to be exact). Europe uses a different video standard, called PAL, which uses 25 fps. Both use a 4×3 aspect ratio but are otherwise incompatible. (If you have friends or relatives overseas, you may have noticed that their videotapes won't play correctly in your VCR—now you know why. There are multi-system decks that play both PAL and NTSC, but most tapes won't work on all types of equipment. The two standards don't even display a frame of video at the same size: a frame of mini DV NTSC video is 720×480 pixels, a frame of mini DV PAL video is 720×576.) In addition to the differences between PAL and NTSC in frame rates and size, there are differences in the ways the frames themselves are constructed. Both video standards use horizontal lines of resolution to create images on a television screen. Images on a television screen may look solid, but they're actually constructed from hundreds of horizontal lines, each of which displays a thin slice of the overall picture. NTSC and PAL both assemble these lines to create images, but they go about the process differently.

NTSC divides each frame of video into two fields, first displaying half the image in a field of odd numbered lines, and then completing the image by displaying the remaining field of even numbered lines (the images refresh so quickly, an audience only notices if something goes wrong). This NTSC method is referred to as *interlaced* video, since each frame is made up of

Muybridge and the Frame-by-Frame Image

At the end of the 19th century, photographer Eadweard Muybridge used still cameras to record a series of motion studies. The collections of still images recorded in quick succession provided detailed frame-by-frame studies of people and animals in various stages of motion that had never before been captured. Muybridge's subjects include, among others, a running horse pulling a carriage, a blacksmith striking an anvil, and a nude woman pouring a bucketful of water onto another nude woman.

These sequentially photographed still images function much like individual frames of film or video, and can be animated by playing them back in succession. Playing back a greater number of images at a higher frame rate produces smoother motion. Playing back fewer images at a slower frame rate yields more choppy results.

HD Aspect Ratio and Frame Rates

Unlike prosumer mini DV cameras, which allow a filmmaker to choose an aspect ratio, high-definition video employs a widescreen aspect ratio of 16x9. The HD format supports a number of different technical specifications and frame sizes, but as a rule, they use widescreen aspect ratio.

Viewing HD video requires a compatible 16x9 monitor that can process a high-definition signal (not all 16x9 monitors can display HD programming). As HD becomes more common, compatibility will become less of an issue. For now, an HD signal can be downconverted to fit a 4x3 frame, a process that entails using a pan & scan effect or letterboxing. Standard definition video (meaning anything other than HD) can also be upconverted to conform to HD standards. This entails scaling a 4x3 frame of SD video to fit the larger 16x9 frame, and leaving blank space at the sides of the screen.

HD also supports a variety of frame rates, including both progressive and interlaced, which are described in this chapter. Frame rates of 60 or 30 frames per second are more television friendly, since they can conform to the NTSC broadcast standard. A rate of 24 frames per second results in easier transfers to film.

two interlocking frames. PAL, on the other hand, uses a single image for each frame of video. This method is referred to as *progressive*, because PAL displays a progressive series of complete images one after the other. Interlaced video frame rates are often referred to in writing by the number of frames per second followed by the letter "i"; for example, 30i refers to 30 fps interlaced video. Progressive video is referred to in writing as the number of frames followed by the letter "p"; for example, 25p. More advanced mini DV camcorders typically list their available frame rates either on the box or in the manual. If you have a camera that doesn't list this information, and you bought the camera in the United States, it most likely records 30 fps interlaced video. Likewise, if you bought a camera in Europe, it most likely records 25 fps progressive video unless otherwise noted.

To complicate things further, film and some forms of high-definition video run at a progressive 24 fps (or 24p)—a full six frames per second less than NTSC video. Making a film print or HD master of material shot on NTSC video requires a computer-assisted *pulldown* process to conform video to the slower frame rate without disrupting the natural motion of the images.

In the past, filmmakers employed a variety of creative workarounds to make their mini DV footage work on the big screen. *Tadpole*, released by Miramax in 2002, was shot and edited in New York using 16x9 PAL equipment because the European format of 25 progressive images per second made for an easier video-to-film transfer than the North American standard of 30 interlaced images. This technique cost significantly less than shooting 35 mm, but using the PAL standard in the United States, instead of NTSC, made postproduction more complicated.

Today, sophisticated and versatile mini DV cameras such as the Panasonic AG-DVX100A offer a choice of frame rates, allowing directors to shoot 30 fps interlaced video for television broadcast or 24 fps progressive video for film or HD transfer. The 24 fps progressive option eliminates the need for pulldowns in video-to-film transfer, and, combined with the camera's ability to shoot 16x9 video, it becomes easier than ever to shoot mini DV and distribute your work in a professional format.

Video on Your Computer, Pixel Aspect Ratio

Unlike video, which uses lines of resolution to create images on a television screen, computer monitors create images using thousands of small squares called *pixels*. When you work with video on a computer, applications such as Final Cut Pro and After Effects display your video as *non-square pixels*. As you can tell from the name, these pixels are shaped differently than the pixels in other computer images. If you're creating images in a program such as Adobe Photoshop, and you plan to use them in a video, you need understand *pixel aspect ratio* to ensure your images don't distort.

Pixel aspect ratio (Figure 2-4) describes the width of a pixel in relation to its height. Square pixels, generated by most computer applications, are equal in height and width —that's what makes them square. Non-square pixels display at different proportions: they appear taller than they are wide. The difference becomes important when you start to combine video and still images. If you create a title or a graphic and then import it into your video, the image may distort slightly due to the difference in pixel aspect ratios. For example, if you create a circle in

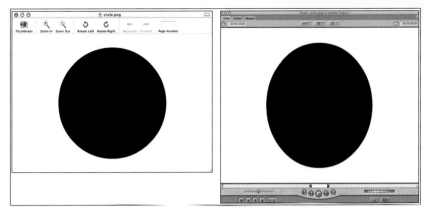

Figure 2-4. In the figure at left, a circle created in a design application looks proportionally correct. At right, the same circle stretches vertically when imported into Final Cut Pro.

Photoshop and then import the circle into Final Cut Pro or After Effects, it might display as an oval. Likewise, if you export a still image of a person from a frame of video, the image can distort on a computer screen making the person appear shorter and fatter.

The key to avoiding distortion is to compensate for different pixel shapes. Unlike the problems that stem from displaying 16×9 material on a 4×3 screen, pixel aspect ratios can be easily managed. Essentially, if you're going from square to non-square pixels (for example, if you're importing a still image into your video project), you can make your images slightly taller than you normally would, and when you import them into a video-production application, they'll resize to their correct proportions. If you're going from non-square to square (let's say you're exporting still images from frames of video), expand the height of the exported image in Photoshop to compensate for the change. Chapter 11 includes specific techniques and details on aspect-ratio management, but I wanted to mention the process here so it's something you're thinking about from the start. Each decision you make in production ripples out to effect later choices and options—especially when you put your work on DVD.

Why It All Matters in DVD Production

DVD production is a great option for independent filmmakers. Until recently, years after digital cameras and editing systems became standard equipment for independents, filmmakers still had no choice but to output their work onto VHS for home viewing. VHS is an analog format, so image and sound quality always suffered. DVDs have been on the scene for a while, but until recently, they cost too much for an individual to produce. Now, computers routinely ship with DVD burners as standard equipment so you can easily output your finished project to disc for less than the cost of making

VHS dubs. To make a good DVD, however, you need to understand both screen and pixel aspect ratios, video standards, and how to manage them to make your work fully accessible to your target audience.

You can make a DVD that displays 4×3 content, 16×9 content, or both. This creative freedom, of course, necessitates a fair amount of planning. For example, you can give the user a choice of watching a letterboxed 4×3 version of your widescreen masterpiece or the original 16×9 version, but you then need to produce a 4×3 version along with a 16×9, and make sure there's room for both on your disc. If your DVD includes widescreen content, you need to define how the material will play on a 4×3 monitor (does the material automatically appear letterboxed on a small screen, pan & scanned, or do you give the viewer a choice).

And don't forget the menus. A former girlfriend used to joke that she would rent DVDs to watch movies and I would rent them to watch the menus (to me, they're often the most creative aspect of DVD production). DVD menus offer almost limitless options, and each option requires thoughtful decisions. DVD menus can contain just about anything you dream up, including moving images, background audio, and complex image montages you create in the editing application of your choice. DVD production applications such as DVD Studio Pro use non-square pixels, so if you're using still images created in Photoshop or backgrounds drawn in Illustrator, menu creation requires careful attention to pixel aspect ratio. In addition to other important factors such as usability and visual design, you still have to remember screen aspect ratio. One of the coolest looking DVDs I ever saw didn't fit on my 4×3 television screen so the edges were chopped off and I had to guess at some of the buttons.

Shooting with DVD compression in mind is covered in the next chapter, and the many benefits of DVD distribution are described in Appendix A. If you have a specific idea of what your DVD should look like, start planning now at the beginning of your project. At this point, you're limited only by your knowledge, imagination, and foresight.

Swing-out Monitor, Viewfinder, or External NTSC Field Monitor

Once you've settled on a frame rate and aspect ratio, you're just about ready to shoot. As I mentioned in Chapter 1, one of the great things about working with digital video is that you see exactly what you get as you're recording. Depending on your shooting style and technical needs, there are number of ways you can go about monitoring your production.

An external NTSC field monitor (Figure 2-5) connected to your camera shows you exactly what your work will look like on screen, so using one is the most reliable way to compose a shot. These monitors are often referred

Figure 2-5. A field monitor, such as the Sony PVM-5041Q, is designed for use on location shoots. It features a variety of professional and prosumer video inputs, and lets you see your footage exactly as it will appear on a television screen.

to as critical monitors, because they don't hide flaws in an image. Unlike consumer televisions, which automatically adjust video signals to make images look better on screen, critical monitors let you see your video exactly as it appears, warts and all, so you can fix potential problems before it's too late. Critical monitors come in various sizes for portable and studio use, but none of them are really pocket sized. If you're doing *guerilla filmmaking*, quickly setting up in one location and immediately moving on to the next, or if you're shooting handheld, a critical monitor's bulk may become an issue. Even the smallest and most portable NTSC field monitor isn't something you can pick up and hold while you're operating the camera, so while it provides tremendous accuracy, it also limits your mobility.

If your project calls for an especially active shooting style, you might want to use a camera with an onboard monitor (Figure 2-6). Many cameras come with a color monitor that swings out from the side of the camera, enabling you to frame a shot while holding the camera away from your body. These swing-out monitors are great for filming in hectic situations where you not only need to see what you're shooting, but you also need to see what's going on around you off camera—for example, if you're chasing someone down the street or filming a complex action scene. Monitors like this also work well with handheld Steadicams (described in detail in Chapter 6), which stabilize a camera but force you to hold it at arm's length.

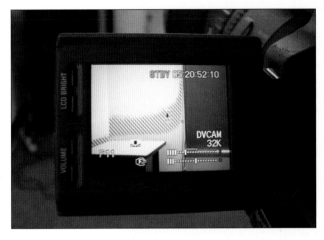

Figure 2-6. If the nature of your shoot doesn't permit you to work with an external field monitor, a camera's onboard LCD monitor may be a better choice. The image displayed in a camera's onboard monitor probably won't provide the exact and precise results of an external NTSC monitor, but the freedom of movement you gain may be worth the trade-off.

A third option is to simply look through the camera's viewfinder. One of the chief benefits of using the viewfinder comes from the way you hold the camera. If you're shooting handheld, the action of holding the camera's eyepiece to your face can help stabilize your shot—the camera, your arm, and your body form a triangular shape that's naturally stable. This is especially helpful when you're using a substantial camera that you can also balance on your shoulder, such as the Cannon XL2. Some directors of photography say smaller mini DV cameras can be difficult to shoot with because their size makes them harder to stabilize. If you hold one of these smaller cameras to your eye to look through the viewfinder, it can go a long way to providing a less shaky image.

Working with a Viewfinder: Color Versus Black and White

The image your audience sees on screen will most likely appear at a substantially larger size than the image you see in a field monitor, and will definitely appear larger than what you see in the viewfinder. Imperfections become more visible in larger images, so while you might not notice soft focus in the viewfinder, it can be really obvious to someone watching your work on a large screen. Viewfinder images with greater contrast make it easier to spot focus mistakes, and black and white viewfinder images generally contain more contrast than color images. So, although black and white may seem really low tech and less attractive, using a black and white viewfinder might help you make a better movie.

Depending on lighting conditions, parts of your shot may be in focus while others aren't. For example, a person close to the camera might be in focus but the pictures on the wall behind them might not. This is referred to as *depth of field*, and skilled cinematographers can use it to draw the audience's attention from one part of the frame to another. Cinematographers can also use focus to make people look more or less attractive—older *Star Trek* episodes invariably feature female lead characters shot in soft focus to make them radiantly beautiful, while evil villains often appear in extra sharp focus so you can see the dark contours of their faces.

Audiences expect images to appear in focus. If a shot appears out of focus, audiences notice. Errors pull audiences out of a story and shift their attention to problems in technical details. As I mentioned earlier, black and white viewfinders display higher levels of contrast, so they may help you avoid focus problems that lessen the technical quality of your project. The same holds true for exposure. If you're using zebra lines to set your aperture (zebra lines, covered in Chapter 4, are black stripes that appear in the viewfinder to indicate overexposed areas of a shot), it's much easier to make adjustments and see the results in a high-contrast black and white viewfinder. Different areas of your frame may require different exposure settings, and spotting exposure problems in the viewfinder enables you to compensate and find a balance. If the viewfinder doesn't display enough contrast, and color viewfinders often don't, it's easy to overlook slight differences in exposure that can become a real problem later on.

Lastly, the color you see in a viewfinder, or even a swing-out monitor, may not be the exact color you see on an NTSC monitor anyway. Color management is a complex and precise process (covered in Chapter 16 of this book) that requires measurement beyond what you can see in a color viewfinder. If you look through the viewfinder and see color that totally doesn't match what you see with your eye, there's a problem that you need to fix. More often than not, however, you won't spot anything in a color monitor that changes your life, so again the black and white viewfinder may be a better option.

As you can see, there's no shortage of details to keep in mind when you're planning a film. The technical concepts introduced in this chapter, including aspect ratio, frame rate, and video standards may at first seem burdensome, but they ultimately provide you with an empowering degree of control. Once you master the basic technical elements of video production (and with time, they become second nature), you can put them to use and

---

NOTE

For a truly detailed explanation of depth of field, including charts that calculate depth of field in various lighting conditions, take a look at the American Cinematographer Manual. *This book provides industrial strength information on a variety of very specific technical issues, and was recently updated (the 9th edition was released in November of 2004).*

---

create a project that looks exactly the way you want it to—and that's what this book is about. Technology is a tool that enables you to realize your creative vision on screen, and the upcoming chapters explore this in increasing levels of depth.

The next chapter examines the mechanics of a shot and explores the art of placing one shot next to another. The chapter opens by explaining the difference between a wide shot, medium shot, and close up (these shots form the building blocks of a director's cinematic vocabulary). The chapter then discusses the effects of juxtaposing different types of shots, and ways you can use the juxtaposition to engage an audience. The chapter also examines the ways some types of shots compress better than others for DVD production, and how you can plan for this when you shoot.

Composing a Shot to Fit Your Output Medium

3

When I was in film school, a professor told us that people learning to use a camera tend to frame things they way they see them with their eyes—everything appears in medium shot. When you look around, your eyes take in everything at once. If you look at a particular object, say a book on a shelf, your eyes don't zoom in and frame out everything around it. Even as you read the words on this page, you can still see objects to your left and right using your peripheral vision.

The medium shot, however, isn't the only option available to a director. As a filmmaker, you have the ability to focus an audience's attention exactly where you want it. Depending on how tightly you frame a shot, your audience might see everything in a room or only a single detail on one specific object. Cinematographers use a vocabulary of different shot types to tell a story and shape their audiences' experience. Directors can also use framing to make a project particularly friendly to a specific viewing environment, such as television, movie theaters, or home screenings via DVD.

An Overview of Shots—Medium Shot, Wide Shot, Close-up, and Extreme Close-up

Medium shots, abbreviated *MS*, are generally the most common type of framing. A typical medium shot might show the protagonist of a film in his environment; for example, a teacher at the front of a classroom. The shot would be wide enough to see the teacher and the space around him, providing a sense of context—you might not know someone is a teacher, but if you see the person in medium shot writing on a blackboard, you can figure it out.

Figure 3-1. Drawn medium shot of a teacher in front of a blackboard.

The medium shot in Figure 3-1 provides the audience with a fairly detailed view of its subject. It allows viewers to get a good sense of what the person in the shot looks like and provides enough context to place the person in a specific environment, in this case, a school classroom.

Wide shots, *WS*, allow the audience to see a large area at once. If the cinematographer decided to keep the teacher in the frame at the front of the room and pull back to include students taking notes, the medium shot mentioned earlier would become a wide shot. Directors often use a wide shot to show multiple actions taking place at the same time, or to let an audience know where a scene is taking place. When a wide shot is used to establish a location, it's called an *establishing shot*.

The wide shot in Figure 3-2 shows the same teacher in the same classroom, but also shows more of the area around her. A wider shot provides the audience with more information about where a scene is taking place. The audience could see from the medium shot that the action was taking place in a classroom, but this wide shot gives viewers a better idea of what the classroom looks like, who the students are, etc.

Figure 3-2. Drawn wide shot of a teacher in front of a blackboard.

A variation of the wide shot is an *extreme wide shot*, or *XWS*, which contains an extremely large area in a single frame. Extreme wide shots make great establishing shots and provide a natural way to open a film. A logical opening shot for a movie about a school would be an XWS of a school building's exterior. Extreme wide shots don't show small architectural details or highlight individual members of a crowd, they show large areas at once. To show detail, such as facial expressions, a *close-up* (*CU*) would be a better choice.

The version of a wide shot in Figure 3-3 is called an extreme wide shot, because it shows such a large area. Extreme wide shots provide a natural way to establish the location of a film sequence. When people in the audience see this shot, they understand that the accompany-

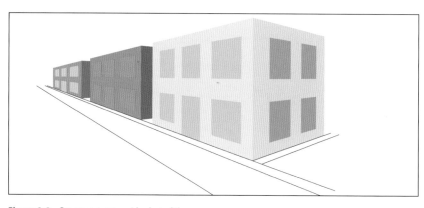

Figure 3-3. Drawn extreme wide shot of the school building, from down the street.

ing medium and wide shots of the teacher are taking place inside this school building.

It's hard to use the phrase close-up without thinking of Norma Desmond in *Sunset Boulevard*. In a state of delusion, the fallen star, speaking to an imaginary director, says "All right, Mr. DeMille. I'm ready for my close-up." (The director is imaginary in that he's not in the scene—the legendary director Cecil B. DeMille actually appears in the film earlier, playing himself. The only cameras at the end of the film, however, belong to news photographers there to film her arrest.) Desmond wanted a close-up, because it meant she was important enough to have the screen all to herself.

To a director the *CU* offers a great opportunity to draw the audience in to the person or object on screen, both visually and emotionally. Want the viewer to fall in love with your star? Show her in warm lighting and frame her face in a full-screen close-up as she smiles softly. Want to scare your viewer? Cut to a tight close-up of something people don't want to look at (*Fear Factor* does this exceptionally well when it's time for contestants to eat things like worms and beetles, or worse).

A good, tight shot brings the audience very close to the image onscreen—both literally and emotionally. If you're a fan of pastrami, Figure 3-4 will likely have tremendous impact. It's a very strong image, in part because the sandwich comes from the famous Katz's Deli in New York City, and if you're ever going to eat pastrami, Katz's is the place to get it. More importantly, the sandwich's proximity to the camera and its dominance of the frame force you to look at the details in the image and take notice. If you're a vegetarian, or otherwise opposed to the consumption of beef, this may be a very unpleasant photo, especially because it's shown as such a tight close-up. Had it been taken in wide or medium shot, it wouldn't have nearly the same impact.

> **NOTE**
>
> *To really show detail you can use an extreme close-up, or ECU. The ECU shows an object in such close detail that people in the audience may have to think for a minute to figure out what they're looking at. An ECU doesn't tell an entire story on its own, it provides striking visual detail and can jar your audience into paying extra close attention. In the opening sequence of* Blade Runner, *director Ridley Scott cuts from panoramic wide shots of an overcrowded Los Angles landscape in the year 2019 to an ECU of a man's eye as he undergoes a scientific test. Several minutes pass before the audience understands what the eye has to do with the plot, and that's a big part of why the opening is successful. The audience spends the opening trying to figure out why Scott included this seemingly random but disturbing close-up, and keenly observes every detail. By the time Scott reveals the significance, the audience is solidly wrapped up in the story.*

Figure 3-4. Extreme close-up of a pastrami sandwich from Katz's Deli.

The 16x9 aspect ratio is especially friendly to wide shots

People who fly into Los Angeles or Mexico City often remark that the crowded landscape of each city extends so far into the distance. Unlike other cities surrounded by open space, LA and Mexico City seem to go on forever. The futuristic LA in *Blade Runner* is even more overpowering, especially on a wide screen. The shape of a widescreen frame allows filmmakers to compose on a large scale, in fact, many films are made specifically for a big screen.

Koyaanisqatsi, Godfrey Reggio's meditation on the imbalance of nature and the man-made world, contrasts breathtaking unpopulated landscapes with mesmerizing shots of crowded urban areas and mechanized workplace environments. Everything in the film appears at a scale far larger than human. The overwhelming nature of the images in *Koyaanisqatsi*, the title of which means "life out of balance," makes a statement, enhanced by the use of widescreen aspect ratio. Aspect ratios of 16×9 or wider enable a filmmaker to work on a large scale.

Badlands, Terrence Malick's 1973 love story about a couple on a killing spree, stars Martin Sheen, Sissy Spacek, and the landscape in which they were filmed—hence, the title. The landscape plays an important role in the story, partially because viewers have become accustomed to watching desperados flee through similar scenery in cowboy movies, but also because the landscape is such a beautiful and serene backdrop for the couple's violence. Like the opening of *Blade Runner*, this film holds up when viewed on a television screen, but its true power and impact comes from its larger-than-life widescreen quality.

CUs and XCUs for the 4x3 screen

Television programs are tailored fit a 4×3 screen, since many people watch them on traditionally shaped analog television sets. Shows that feature interviews, such as *48 Hours* or *60 Minutes*, often show people in close-up shots that frame out everything but a person's head and shoulders. Rather than attempting to shrink a large area of action to fit a small screen, television producers select a limited area of action and focus attention on it. For example, police shows often feature chase scenes, but the action isn't shown as a panorama. Even wide shots focus the viewers' attention on the characters as opposed to the environment or landscape.

Television programs also make great use of ECUs. The opening sequence of *The Odd Couple,* which premiered in 1970, shows an ECU of a frying pan handed to Felix Unger as his wife kicks him out, asking that he never return, followed by an ECU of Unger's index finger extended toward the doorbell of his childhood friend Oscar Madison. These opening shots set the stage for a series about two divorced men who live together, and they make great use of the 4×3 screen. Even on the smallest television screen, audiences can still clearly identify frying pans and doorbells. The panoramic details of *Koyaanisqatsi* aren't made for the small screen, instead, the extreme close-up is a perfect fit for TV.

Home viewing has also left its mark on Hollywood. When you go to the movies, you may notice fewer shots of broad landscapes and more close-up faces. Before the influence of cable TV and home video rentals, close-up shots were reserved only for the big star, but now they're being used with greater frequency on secondary characters, because they fit a television screen so nicely.

Editing, the art of placing one shot next to another

Koyaanisqatsi contains a strikingly enigmatic close-up of a woman at work. The face of a woman, wearing glasses, fills the screen, and the director holds the shot. It would be entirely unremarkable, except for a reflection moving across her glasses. The director keeps the shot onscreen long enough to make people wonder what the moving objects might be—the rest of the shot is eerily still, but the reflection in her eyeglasses keeps moving. The next shot reveals the woman working in a mail sorting facility, and the audience discovers the reflections are envelopes. The viewer's curiosity is piqued and then satisfied. If these shots had been shown in reverse order, with the wider establishing shot shown first, the sequence would lose its dramatic visual tension.

Placing one shot next to another is the elemental process of building a film. Like everything else in film production, editing is both a technical and artistic activity. The content of each successive shot determines how the audience will see your story unfold, and the framing of each shot helps determine how well one shot works next to another. The previously described *Koyaanisqatsi* sequence works so well because it uses different shot types to create a dramatic effect. The close-up provides the viewer with much less context than the wide shot, so they both serve very different roles within the sequence. The clear difference in framing also makes them compatible.

Regardless of screen size and aspect ratio, working with a variety of shot types makes editing easier. Cutting from a close-up to a wide shot provides a natural transition—the wide shot places the close-up in context. Likewise, going from a wide shot to a close-up instantly makes sense to a viewer, because the CU shows added detail (directors often use this combination to show a person's facial expression in a *reaction shot*). Going from a medium shot to another medium shot, however, doesn't always work. In fact, editing two overly similar shots together can look like a mistake.

The jump cut, cutaways, and coverage

If you took a continuous shot of a person walking and cut a few seconds from the middle of the shot, there would be an obvious gap in the action and you would be left with two short clips that don't work together. At the point in the clip where you made an edit, the motion of the person walking would no longer look smooth—the person would appear to jump. This is referred to as a *jump cut*. It jars the audience out of the flow of your work and alerts them to the fact action is missing.

Figure 3-5. The top sequence shows an obvious jump cut. The bottom sequence uses a cutaway frame to mask the edit.

The sequence of images in Figure 3-5 represents a jump cut. Images have been removed from the middle of the sequence, so the boy jumps from one side of the frame to the other instead of moving across incrementally.

In the bottom sequence, a cutaway frame has been inserted to mask the point of the jump cut (in this case, the image is a picture of the boy's mother). This distracts the audience from the edit and smoothes the transition from one part of an edited sequence to the next.

This doesn't mean you have to leave long, boring shots in your movie. Editors often trim the beginning and end of a shot, or keep the beginning and end of a shot intact and cut out the middle. To avoid jump cuts, editors use a *cutaway* shot to cover the edit point. A cutaway is, literally, a shot that cuts away to another subject. For example, if a television news editor wants to shorten a long-winded politician's announcement, she can show the politician speaking on camera, briefly cut to a crowd of people listening, and then cut back to the politician. The editor may have eliminated a substantial portion of the politician's speech, but if she makes effective use of the cutaway, the audience will never know.

To avoid the use of jump cuts, a filmmaker shoots *coverage*. Coverage ensures you get shots at a variety of framings (close-up, medium, wide, etc.) and from different angles. If you cut from a shot of a person facing the camera to a shot of that person's profile, the edit is not a jump cut but shows a different point of view. During the production of *Metallica: Some Kind of Monster*, the filmmakers often shot with two cameras simultaneously, both to ensure they didn't miss any action, and to provide the coverage they needed for editing.

Continuity and detail management

In my junior year of college, I had a writing professor who said, "God is in the details." The next semester, I had another professor who told us, "The devil is in the details." Regardless of how you look at it, details can make or break your project—especially if the details are not consistent from one shot to another. If you show two people on screen together in conversation, and then cut to a close-up reaction shot, does everything match? If you film the two shots outdoors at different times of day—for example, you shoot the *master shot* of both people talking at noon and the CU reaction shot at 4 p.m.—shadows on the actors' faces might not match because the position of the sun has changed.

A clock on the wall may show hours have passed between shots, while only a few minutes have passed in your story. An actor might take her hat off in one shot, but have the hat on her head in the next. There are entire web sites dedicated to pointing out continuity errors in films (as I wrote this section, I typed "continuity errors" into Google and got more than 26,000 results).

S I D E B A R

The Rule of 180, or Crossing the Axis

As a camera films action from a variety of angles, it's helpful to the audience to keep the camera on one side of an imaginary 180-degree axis. Shooting from a variety of angles on the same side of the axis gives an editor lots of good options to work with. Crossing the axis can be very confusing to an audience. If you shoot people moving across the screen, as in the image below left, the audience expects them to keep moving in the same direction. If you then show the same people from a reverse angle, as in the image below right, they'll appear to be running in the opposite direction. The motion of the people on screen has not changed, but it looks to the audience as if it has because of the camera angle.

In addition to the arrangement of objects in front of a camera, the location of the camera itself can shape the continuity of a sequence. Although showing everything from the same angle becomes monotonous, too many contrasting angles can confuse an audience. It's become a convention in modern filmmaking that you can safely shoot along a 180-degree axis—an idea often explained in terms of a football game. To a camera on one side of the field, everyone might appear to be running from left to right. If you show the exact same action through a camera on the opposite side of the field, the same players would appear to be running in the opposite direction, from right to left. If you exclusively use footage from either one of these cameras, there won't be any problem. If you cross the axis and mix shots from both sides, it becomes very hard to follow the game.

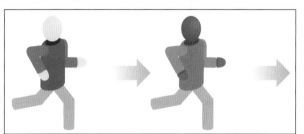

Continuity is such an issue that large-scale film productions hire a *script supervisor* to ensure that continuity problems don't come up.

Script supervisors often take Polaroid or digital photos to keep track of exactly what people were wearing in each shot, how furniture was arranged, where food was placed on the table, and any other details that might get lost between the end of one shoot and the beginning of the next.

Timing

As a filmmaker, your editing style sets the tone of your project. The timing of each edit and the pace of each shot are determined not only by your film's content but the emotional impact you'd like your film to achieve. Quicker cuts create a more exciting tempo, while longer shots establish a more thoughtful, contemplative tone. Editing implies a passage of time, and as a director, you get to decide how much time passes in any part of your show.

For example, you can cut from a medium shot of a person putting a key into a lock to a matching ECU of the key turning (this is called a *match edit*). Alternatively, you could show a person laying his head on a pillow at night and omit hours from your story by cutting to a shot of the person waking up the next day (you could also show the person in bed for a really long time—Andy Warhol once made an eight-hour film about a man sleeping). Quentin Tarantino opens his 1994 film *Pulp Fiction* with a sequence that takes place at the end of the movie. As the director, you're the captain of the ship, and controlling the passage of time gives you complete control of your story.

Shooting with DVD Compression in Mind

As I mentioned in the last chapter, DVD production offers fantastic opportunities for the independent producer. Using prosumer equipment, you can affordably shoot, edit, and distribute professional-quality copies of your work in an all-digital format. Editing applications and DVD production packages ship with encoders that convert your DV footage to the DVD-compatible *MPEG-2* format. Understanding how these encoders function before you go out and shoot can help you create better quality DVDs.

Not all shots compress as easily, or as well, as you might want them to. Certain backgrounds and some types of movement might look great when you record them but lose significant quality when compressed for DVD. Once you develop a solid understanding of DVD compression, you can compose shots that still look really good after they've been compressed.

As always, your artistic vision is the most important factor. If you're spending the time, money, and effort to make a film, you have to include the shots you want and the images you think people want to see—even if they aren't the most compression friendly. Understanding DVD compression can help you ensure your most ambitious compositions and framings translate to DVD the way you want, with a minimal amount of headache.

Compression algorithms

Audio and video files take up tremendous amounts of memory. Five minutes of DV footage takes up approximately 1 GB of memory on your computer's hard drive. A single DVD only holds about 4.7 GB, which translates to less than 25 minutes of *camera-format* DV (which means DV that's been captured from a camera without applying any additional compression). Just about any disc you pick up at the video store plays for more than 25 minutes, and the reason producers can fit more material onto commercial DVDs is the video and audio have been *compressed*. Compression removes redundant information from a digital signal, which results in a more manageable file size. DVD compression programs, such as Apple's Compressor, use algorithms to analyze digital audio and video and identify information that can be removed without anyone noticing. These algorithms (which are essentially complex mathematic formulas) use two types of compression, *spatial* and *temporal*.

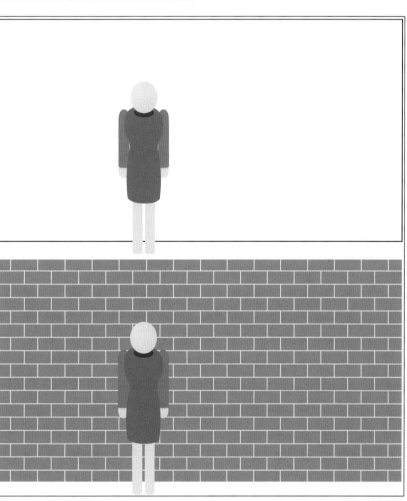

The image on top in Figure 3-6 lends itself to easy compression, while the image below it does not. The image of the person standing against a white wall compresses without difficulty, because much of the image, specifically the white wall, contains very little detail. The image of a person standing in front of a textured background is more complicated, therefore more difficult to compress. While good compression is still possible, the results may be more noticeable. The background contains fine areas of detail that make it harder to simplify.

Figure 3-6. Two drawn images: one of a person standing in front of a white wall, other of a person standing in front of a finely textured background.

Spatial compression compares different parts of a single frame of video to determine what information repeats in the image. Spatial compression algorithms break an image down into blocks, and if parts of a frame repeat, the algorithm places the same block in more than one location. If, for example, a person is standing in front of a white wall, the compression algorithm repeats one section of the background in multiple locations to recreate the wall, rather than saving an image of the entire wall, which would require more memory.

If you've saved an image for the World Wide Web as a *JPEG* (also written *jpg*, pronounced "Jay Peg"), you've used a form of spatial compression. JPEG compression examines similarities in an image, and, depending on the quality setting, eliminates higher or lower amounts of information. Spatial compression of video clips works essentially the same way: higher compression settings result in video that requires less storage space. As you apply increased amounts of compression, you also lose image quality.

In Figure 3-7, although each frame in the sequence is different, the background remains a constant. Even though the person walking is in a different position in each frame, the background is still a solid white wall. This enables the compression algorithm to redraw areas of the wall over several frames, instead of creating an entirely new background for each frame of video. This saves space on the disc and reduces the processing power needed to play the compressed finished product.

If material is too difficult to compress, or if a sequence contains too much information for a DVD player to process during playback, the video will break down. Symptoms include large blocks appearing in seemingly random locations on the screen, or the DVD player may momentarily "hang up" or pause during playback.

Figure 3-7. A series of drawn figures showing a person walking past a white wall.

Temporal compression compares multiple frames of video to look for information that repeats from one frame to another. If we use the same example of a person standing in front of a white wall, the wall doesn't change from frame to frame. Rather than saving multiple frames of the unchanged wall, temporal compression would use a single image of the wall repeated over several frames to save information. Algorithms that use both spatial and temporal compression would save information by repeating small sections of the wall over several frames.

Shots that make compression easier

Evenly lit, solid backgrounds make compression much easier. Finely detailed backgrounds, such as textured drapes, are harder to compress, because the detail requires more information. I once worked with a group of students who shot some video in front of a cork bulletin board. It looked great until they tried to compress it. The small differences in color made compression difficult, and the cork looked like a blocky mosaic in the compressed version. You may notice similar problems if you compress video of people wearing striped or patterned shirts—algorithms often have trouble processing fine details that appear close together.

Pans and tilts also make temporal compression more difficult. Temporal compression compares one frame of video to the next, so if the contents of each frame are different due to camera movement, the material becomes harder to compress. If the camera doesn't move, much of the background will likely be the same from one frame to another, even if there's movement in the foreground. When the camera moves, everything is slightly different in each frame, because the camera records each frame from a different angle. Handheld camera work by definition makes compression more difficult because of the motion—even the most experienced and steady camera operator doesn't stand perfectly still when she holds a camera. Mounting the camera on a tripod creates a more stable shot, which is easier to compress.

For the same reasons that camera movement makes temporal compression more difficult, background motion does also. If the background of your shot contains motion—for example, textured drapes blowing in the breeze, leaves rustling in a tree, or reflected sunlight sparkling on water—the shot becomes more difficult to compress.

The easiest footage to compress would be a static shot of a solid wall, with nothing happening in front of it. This would also, of course, be a very boring shot that not too many people would watch. When it comes down to it, you have to include the action that makes your film exciting and the backgrounds that make each shot worth looking at—even if they don't compress all that easily. This doesn't mean you should ignore the realities of compression when framing your shots, if you shoot actors in front of the wrong background, compression may become a frustrating experience, especially if the shot involves a pan or a tilt.

Before you assemble your full cast and crew, choose the locations you want, record a few test shots, and then compress some of your footage—you can't tell what a shot will look like until you actually see it on a screen. You may find out everything compresses exactly the way you'd like, or you may decide to make some changes in your composition. Either way, your goal as a director is to focus an audience's attention on your story, not on technical problems. Try a variety of framings, and incorporate a number of different

> **NOTE**
>
> *In situations where a shoot requires more mobility than a tripod allows, you can use a camera stabilization device such as a Steadicam or Glidecam. Both are described in Chapter 6.*

camera angles and backgrounds. This way, you'll be all set when your star tells you she's ready for her close-up.

This chapter introduced different types of shots, the ways they work together, and the idea of juxtaposing a series of different shot types to form a sequence and tell your story. This chapter also explored the way different compositions, visual elements, and framing choices impact the way footage compresses for distribution via DVD (the same principles apply if you stream video over the Internet, which this book explores in Appendix A).

As a director, you have to have a mind that is good at details as well as the bigger picture. If you have the mind for both, you're going to be a good director. If you don't have the mind for both, pick people for your crew that have the mind for the aspects you lack.

The next chapter addresses lighting, which can make your work look very polished and professional if handled well, and is immediately noticeable to an audience if there are any problems. The chapter covers ways to light various situations, and why you should always use the manual settings instead of relying on a camera to automatically make adjustments for you (as I always tell students, talented people are always much more valuable to a production than any piece of equipment on its own—that's what enables you and me to make a living).

NOTE

It's important to know about technical details, but don't let them limit your creativity. During the planning stage of my most recent project, Mi Querdia America, *I spent a great deal of time discussing compression strategies with my director of photography. The project follows a group of immigrant teens through their first year of high school in New York City and was shot to be available entirely online. As you can imagine, maintaining good visual quality after compression was a top a priority. After some very lengthy discussion, I finally looked at my DP and said, "You know what, just make it look good." He did—the video he shot was great, and I worked out the compression details later. The results are online at http:// www.digitaldocumentary.org/ america.*

Lighting for Digital

4

At the most fundamental level, lighting allows an audience to see what you're shooting. Digital cameras function much like the human eye, transforming different amounts of reflected light into recognizable images. To a skilled filmmaker, lighting is also a tool to add depth, texture, and nuance to a composition. Careful use of shadows, color tints, and highlights can transform an otherwise unimpressive shot into something that speaks to your audience on both aesthetic and emotional levels. Skillful lighting can make a good shot great and a great shot epic.

What White Balance Does, and Why You Can Never Forget to Set It

Most lighting isn't neutral. Most light contains a *color cast* that tints the objects in a shot. Even if you can't see the tint with your eye, it makes a difference to the camera. Depending on lighting conditions, the light in a location may have a blue or a yellow tint that colors everything you see. The human eye and brain do a remarkable job of compensating for differences in lighting, so most of the time people don't even notice a tint unless it's really strong. If you stand next to a red light bulb, you'll notice it's turning everything around you red, but most people don't notice that everyday light bulbs make things look yellow. Cameras, however, don't compensate—they record the lighting conditions in an environment, for better or worse.

In contrast to a light bulb's yellow tint, daylight looks blue. You can sometimes see the difference in *mixed lighting* situations—if you turn on a lamp at dusk, you may notice the natural light coming in through the windows makes the room look blue, except for the area around the lamp, which looks yellow. Cinematographers describe the difference in terms of *color temperature*. The concept refers to the range of color a light source emits, and is measured as a temperature in units called Kelvin. At the lower end of the scale, light sources emit more red and yellow light. At the higher end, sources emit more blue light. Daylight at noon on a sunny day measures approximately 5500 k, and tungsten light measures approximately 3400 k. (Tungsten refers to the metal used in the filament in a specific type of bulb used in film and video lighting; other kinds of indoor lighting, such as fluorescents, produce different color temperatures and different lighting tints.)

Each of the three lighting types pictured in Figure 4-1 (from top to bottom: fluorescents, commonly available household light bulbs, and tungsten production lights) creates a different color cast. In mixed lighting situations, they can be notoriously difficult to work with in combination. If you're shooting in an indoor location with mixed lighting and your subject moves across the room, he will appear to change color when moving from one area of light to the next. This might look really cool if you're shooting a music video in a disco, but other than that, it's a good thing to avoid.

To avoid problems with mixed lighting, filmmakers often turn off all the household light bulbs and any fluorescent lights in a room and rely on their own tungsten lamps to light a shot. Turning off fluorescent lights is an especially good idea, because in addition to making people look green in video, fluorescent lights flicker. The flicker might not be visible to the human eye, but in a video recording, it can look like a strobe effect. Again, if you're going for a disco look, you're all set; otherwise, turning off fluorescents is a good habit to get into.

To compensate for differences in color temperature, reels of film are *balanced* to shoot in daylight or tungsten environments. A filmmaker shooting outside would use a reel of daylight film, and would switch to a reel of tungsten for an indoor shoot. Use the wrong type of film in the wrong lighting, and you're stuck with discolored images. DV cameras, as you may have noticed, use the same tape regardless of lighting conditions. To compensate for differences in lighting, DV camcorders have a *white balance* setting.

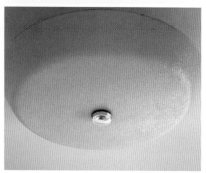

Figure 4-1. Photos of a fluorescent light fixture, a fixture that uses a household light bulb, and a tungsten film/video light.

DV camcorders adjust their color by taking a sample of white area or object in a scene each time the camcorder setup is changed. Because lighting varies from location to location, setting an accurate white value becomes especially important—a bad white balance throws off all the other colors in your image. If you ever looked at a tape and noticed that all the images looked especially yellow or blue, the white balance was probably set incorrectly. Some experienced directors intentionally use a distorted white balance to create a desired effect, most of the time a bad white balance simply means someone behind the camera made a mistake.

The color temperature of indoor lighting varies depending on the type of lights in use, and color temperature outdoors can vary by the time of day (dusk and dawn are lower on the scale than midday). It can even vary in a single scene when the sun goes behind a cloud. Many cameras offer preset white balance modes for outdoor, indoor, and nighttime shots, but the safest way to set a truly accurate white balance is to use your camera's manual white balance function. If you're using lights in an indoor location, daylight coming into the room through any windows can create a mixed lighting situation that renders a preset white balance mode useless anyway.

Setting a manual white balance varies from camcorder to camcorder, but the essential action involves zooming in on a flat white object so it fills the frame (you can buy expensive white balance cards or use a white loose-leaf binder, white foam core, illustration board, or even a blank piece of paper) and then pressing a button to set the white balance. This manual function specifically defines what white looks like in a particular lighting situation, and the camera adjusts all the other colors accordingly.

For example, to set the white balance on a Sony PD-170, as shown in Figure 4-2:

1. Fill the frame with a white object (make sure it's white and not off-white or light gray).

2. Press the button labeled "WHT BAL" on the back of the camera.

3. Press the SEL/PUSH EXEC wheel. The white balance icon in the display (or viewfinder) will begin to flash.

4. Hold the SEL/PUSH EXEC wheel until the icon stops flashing.

Figure 4-2. A Sony PD-170 from the rear.

When you're shooting indoors, a window becomes a light source. Daylight, or natural light, can be very bright—even when you're inside. Filmmakers often use natural light to their advantage, especially when composing interview shots. If you seat a person near a window, facing the camera, you can use natural light to create an attractive highlight on the side of her face. The parts of a person's face farther away from the window will fall into shadow, creating a very nice contour effect. If the shadows become too dark, you can set up a light or reflect some daylight to fill them in. Even if using daylight creates a mixed lighting situation, you'll be okay as long as you set the white balance and as long as your subject stays in the same part of the room.

The Importance of Setting an Aperture

Exposure is a film term describing the way light interacts with a negative to create an image. Too much or too little light produces images that are either too dark or too light. In DV camcorders, exposure works in a very similar way. Depending on the *aperture* you set, greater or lesser amounts of light reach the sensors that produce an image in a DV camera (if you have a digital still camera, it works the same way). Wider apertures let in more light, and smaller apertures allow less light to pass. Prosumer DV cameras allow cinematographers to manually set an aperture, rather than

NOTE

The key to using zebra lines is to adjust your exposure so the lines are just barely visible in the brightest areas. If you can see zebra lines all over your frame, your aperture is set to allow too much light. If you don't see any lines at all, the exposure may be darker than it needs to be. If you open up the aperture until you see lots of zebra lines and then gently stop down, or close the aperture, until they begin disappear, the point at which they're just barely visible will provide the exposure you want.

rely on an automatic setting. To some people, manually setting the aperture may seem like extra work—to a professional, manual exposure settings provide precise control over one of the most important camera settings. (Racecar drivers don't have automatic transmissions under the hood at the Indianapolis 500.) Even if you're not shooting the most important shot of your life, manual aperture settings provide valuable control. The aperture of a camcorder functions much like the iris in a human eye—both expand to let more light in or contract to keep light out. For this reason, filmmakers often say "iris up," to capture more light in a darker environment, or "iris down," to close a camcorder's aperture setting in a bright area.

As I mentioned in Chapter 2, different parts of a frame often require different exposure settings. If you let the camcorder set the aperture for you, you have no control over which part of the frame is properly exposed and which is not. If you set the aperture yourself, you increase the available possibilities. For example, you can film someone standing in front of a bright area and set the aperture so he is properly exposed, while the area behind him appears as an abstract image. Alternatively, you can expose for the bright area behind the person, and the person will appear in silhouette. Depending on the situation, it may also be possible for you to split the difference and choose an aperture that properly exposes the foreground and background of your composition.

Prosumer camcorders offer the option of zebra lines that display a visual warning in overly bright parts of a frame, as shown in Figure 4-3. For example, if you're shooting an actor outdoors against a blue sky on a sunny day, and properly expose for the actor, the sky will likely appear *blown out*, or over exposed. A camcorder set to display zebra lines would show thin, diagonal lines to indicate a problem. You could then either recompose your framing, or *stop down* and choose a smaller aperture, which lets in less light. When the zebra lines disappear, it means you're good to go. If, instead, you let the camcorder automatically choose an exposure setting for you, you narrow the range of possible shots.

Figure 4-3. The diagonal green lines in this image are called zebra lines, and indicate areas of the frame that are too bright (not surprisingly, it's the area around the lights).

Indoor lighting

Careful lighting and subtle shadows give an image texture and depth, but many video images are painfully flat. Sometimes filmmakers take such pains to evenly expose all areas of a frame, especially when lighting interviews, that a shot loses all its shading and nuance. I once read an interview with a restaurant owner who said the only thing worse than a restaurant that's too loud is a restaurant that's too quiet, because it has no customers. I have similar thoughts about video images—out-of-control shadows and areas of impenetrable darkness are no good, but a complete absence of shadows isn't the greatest thing either. (Keep in mind that after you learn the rules here, you might want to occasionally break them for an effect. Sometimes—not often, but sometimes—you might *want* out-of-control shadows and areas of impenetrable darkness. Especially if you're shooting a horror film; see the discussion of darkness in *Open Water* in Chapter 16.)

Working with lights

More often than not, the lighting in a particular environment may not be bright enough to accommodate a video shoot. In that case, you'll either need to buy or rent a light kit. Each lighting kit is configured differently, depending on the kinds of lights and stands you're using, but they all follow some basic principles. All lights fold up for storage and transport, so the first step in lighting a shoot is taking your lights out of the kit and setting them up.

Opening a light stand

The base of a light stand is, essentially, a tripod that folds up like an umbrella. The light stand in Figure 4-4 is shown in the open position. When the legs are extended into the widest position possible, the light stand is at its most stable.

To open a closed stand, slide the part labeled A down until the legs open to their widest possible position, as shown in Figure 4-4. This stand is now ready to use. (To close it, slide the part labeled A up until the legs fold in flush against the body of the stand.)

If you need to raise the light higher, loosen the screw that releases the lowest telescoping section (labeled B in Figure 4-4), extend the height of the stand, and retighten the screw to hold the stand in its extended position. If you need even more height, extend the next highest section.

> **NOTE**
>
> *Be sure to tighten the screw well enough that the stand will not collapse when you place a light on top, but don't over tighten because you can damage the threads in the screw.*
>
> *Always extend a light stand from the bottom section up. The lower telescopic sections are thicker and more stable, so they can safely support more weight.*

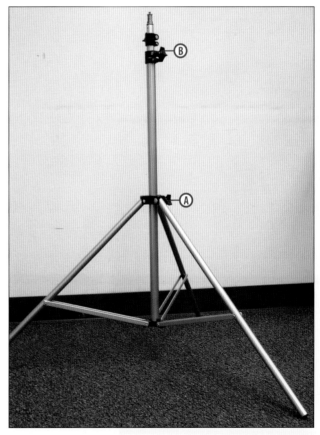

Figure 4-4. An open light stand.

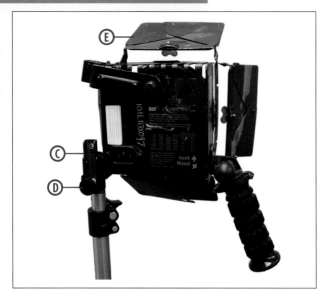

Figure 4-5. A light mounted to a stand, shown from the rear.

Figure 4-6. A light with the power cable connected.

Placing a light on a stand

Once you have adjusted the stand to your desired height, you can add the light itself. Figure 4-5 shows how to do this.

Lights contain a housing (labeled C in Figure 4-5) that fits over the top of a light stand. The housing contains a screw that holds the light in place (D). Loosen the screw until there's enough room to comfortably fit the housing over the top of the light stand. When the light stand has been mounted on the stand, retighten the screw so that the light is securely attached to the stand.

The metal flaps hinged to the edges of the light (E) are called *barn doors*. When they're open, more light comes out; when they're closed, less light comes out. Operating a light with the barn doors completely closed can very easily cause the light to catch fire—always open the doors before turning on the light.

Next, connect the power cable, as shown in Figure 4-6. Most often, a light's power cables are not permanently attached. The light pictured here has a power cable (F) that completely detaches during storage. Always connect the power cable to the light before plugging it into the power source.

Now you can safely plug the light into a power source. Production lights use large amounts of electricity, so try to avoid plugging more than one light into the same outlet—you can easily blow a fuse or trip a circuit breaker. Depending on the lights you're working with, some will turn on as soon as you plug them in; others will have an on/off switch.

Fine-tuning your light set-up

The keys to interview lighting are diffusion and light source placement. In film terms, *diffusion* is a fabric made of spun glass that goes in front of a light to soften the output. Diffusion doesn't change the color of the light, it just makes the light softer and more manageable to a cinematographer. If you place a person in front of a light source without any type of diffusion, the light simultaneously creates very bright areas and very dark shadows. The same light source with a sheet of diffusion applied creates a more even light: the bright spots appear as pleasant highlights, and the dark areas show

---

SIDEBAR

Working with Lights: Dos and Don'ts

Following are some things to keep in mind when working with lights:

- Keep the lights turned off until you need them to shoot or to test your lighting. As soon as you turn production lights on, they get very hot. This can quickly make the room you're in uncomfortably warm, and it makes the lights impossible to handle.

- Turn the lights off as soon as you're done shooting (and give them time to cool off before you try to handle them).

- Never touch the bulb of a production light with your bare fingers, even when the light is off and cool. Your fingers can leave traces of oil on the bulb that will cause it to explode when it heats up.

- Never set up a light under a sprinkler. Sprinklers function by detecting heat. Set up a light or two underneath, and your shoot will probably end earlier than you want it to—and not in a way you'd like.

---

detail and contour while remaining light enough for a cinematographer to work with. The downside to using this type of diffusion is that lights get very hot, and sheets of diffusion can catch fire—not what you want during a shoot.

Until recently, filmmakers went to great lengths to produce diffuse lighting effects—for example, bouncing a light off a reflective umbrella or placing squares of spun glass material over the face of a light (both of which can easily cause a fire if you're not careful). Around the mid-1990s, lighting companies began introducing soft lighting kits, such as the Chimera, made by Chimera Lighting (shown in Figure 4-7), or the Lowell Rifa-lite, which are basically tents that enclose a light and have diffusion built into a front panel. The diffuse light of the Chimera allows a cinematographer to create intricately sculpted lighting effects that fade from light to dark under the control of the filmmaker. The goal of the cinematographer may be to create strikingly eye-catching lighting scenarios, or to subtly highlight a person's facial features and gently capture the audience's attention by drawing a viewer

Figure 4-7. Using a diffuse light source, such as a Chimera, enables you to create a soft, even light.

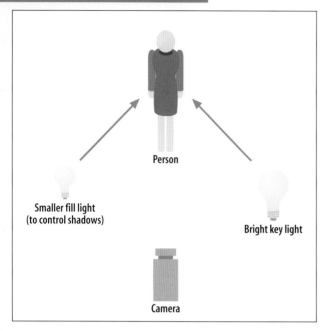

Figure 4-8. The main, or "key," light source is very often a diffuse light such as a Chimera or similar type of production light. The smaller "fill" light is used to fill in areas of shadow that might otherwise get too dark.

in to the person on screen. Good control of lighting enables a number of approaches.

To use a Chimera to greatest effect, filmmakers often place one at an angle, pointed at the face of the person being filmed. Placing the Chimera at an angle helps make sure the person on camera doesn't have to stare directly into a light source (your actors will thank you). It also maximizes the light's ability to generate highlights and shadows, creating pleasant and attractive contour effects. To make sure the side farthest from the Chimera doesn't become overly dark, a filmmaker would likely place another light on the subject from another angle. For example, if a person is looking into the camera, I might use a Chimera to light them from one side and a smaller light source to light them from the other. The smaller light is used to fill in any unintentional shadows, and for that reason, it is called a *fill light*. The Chimera serves as my primary light, creating highlights, and is called a *key light*. This method is called the *key and fill* light combination. Placing both lights in the right spots ensures you don't get any shadows that are overly dark or unmanageable, and also creates nice highlights so your video doesn't look too flat. See Figure 7-8.

The next time you watch a well-produced interview show on television, check out the lighting. You may notice the techniques described in this chapter. More than likely, you've seen them before—audiences generally don't notice good lighting on a conscious level, it's just something they expect to see. Bad lighting, of course, stands out instantly, indoors or out.

> **NOTE**
>
> *If you're working in a controlled interview situation, and you want to make a shoot look really sharp, you can add a hair light. A hair light adds a nice, crescent-shaped shine to the head of a person being interviewed. The shine not only makes your interviewee look attractive and healthy, but it helps differentiate the edge of her head from the background, making her the clear focus of your frame. Filmmakers go to great lengths to light their interviews, because the better an interview looks, the more effective it is in a film. For a very nice example of a hair light, go to www.digitaldocumentary.org/america and click on the clip labeled "Christmas," or the one labeled "bathroom" (don't worry; it's nothing inappropriate).*

Outdoor lighting

People learning to shoot film or video often think that bright, sunny days are the best time to film outdoors. Although bright, sunny days are great times to be outside, they don't offer the best shooting environment—bright sunshine creates dark shadows. If your framing contains areas of bright sun and dark shadow in the same shot, say a sidewalk partially shaded by a tree, it's hard to expose for both. If your shot involves action that moves from the shadowed area into the sun, it becomes even harder to film. Bright sunshine also creates bright reflections. If your shot contains reflected sunlight in windows or on shiny cars, these areas may be *blown out* if you expose for the rest of the frame.

Overcast days, on the other hand, offer nice, even lighting and very few shadows. Even very cloudy days offer ample sunlight for shooting, while avoiding the problems of glare and unworkable shadow. Unfortunately, not even the most talented director has control of the weather. The solution? A *neutral density (ND) filter* and a *bounce card*.

An ND filter is a glass filter you place over the lens of your camcorder (some prosumer cameras offer a built-in ND function; check your manual). An ND filter doesn't change the color cast of your shot, that's why it's called neutral density, it just makes things darker. If you're filming on a very sunny day, it may be too bright to get a good exposure, even at the smallest aperture setting. Using an ND filter evenly darkens the objects in your shot to make them more workable (I once went out for brunch on a cloudy day and a friend remarked that the weather looked like a perfect ND filter). Because an ND filter doesn't change the color of your shots, the audience will never know you're using one—this filter simply gives you the ability to capture images in a wider variety of settings.Once you get exposure under control, you can then even out shadows using a *bounce card*. If you live in a sunny region, you may have a folding reflective screen you place in the windshield of your car to keep the inside from getting too hot when you park in the sun. A bounce card works the same way, it reflects sunlight, and on a film shoot, you can use one to redirect sunlight where you want it. The same way you use a key and fill light combination indoors, you can work with natural sunlight and a bounce card to get the lighting effect you want. You don't have to go out and buy an expensive product to do this; you can use the same type of reflector sunbathers use to reflect light onto their faces, or you can even use the reflector from your car windshield (I've done it myself)—even if your car gets hot, it's worth it if it makes your film better. Compose the shot you want, and if you notice shadows that are too dark to work with, use a reflector to bounce light where you need it.

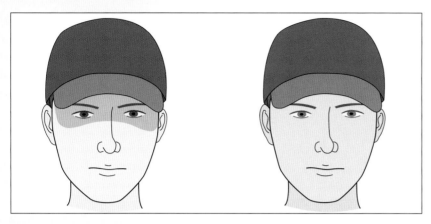

Figure 4-9. Here are two drawn figures of a person wearing a baseball cap. In the first figure, the camcorder exposes for the dark part of the person's face, shaded by the brim, so the rest of the image is blown out. In the second figure, the entire image is more evenly exposed.

Another situation in which an ND filter and a bounce card come in handy is when filming actors in wide-brimmed hats and baseball caps, which can be very hard to work with on camera because they create areas of dark shade. In Figure 4-9, the image at left was shot on a sunny day, so the shaded area of the person's face is significantly darker than the rest of the frame. If the cinematographer exposes for the person's face, the rest of the frame will be too bright. If the cinematographer exposes to keep the majority of the frame usable, the area around the person's eyes will be too dark to see, as in Figure 4-9. In the image at right, the cinematographer has applied an ND filter, making the whole image darker, and positioned a bounce card under the person's face to redirect light where it's needed and lessen the impact of the shadows.

As a director, you always have the option of asking someone to remove his hat. If you think the hat adds to the shot, or if you feel uncomfortable asking, an ND filter and a Flex-fill are good tools to have at the ready.

Shooting outside at night presents an entirely different set of issues. As opposed to midday shoots on a sunny day, where a director has to worry about too much light, a filmmaker shooting at night has to make sure there's enough light to get a decent exposure. A good workaround technique is to shoot at the end of the day, just after sunset. There's a time at dusk, between sunset and the full arrival of nighttime, when there's still enough daylight to shoot. Because daylight looks especially blue at this time of day (take a look next time you're outside just after sunset), it looks like night on screen. If your shot includes any traffic, the effect is reinforced by drivers who turn on their headlights—the light from the cars won't change your exposure, but when audiences see cars with their lights on, it further establishes the shot as taking place at night.

The time immediately before and after a sunset is referred to as *golden hour*, because the setting sun gives off an especially warm glow. This glow looks great onscreen: landscapes look scenic, and people look healthy and attractive. Many directors intentionally wait until the end of the day and schedule shots at golden hour to make their lead actors look especially good. Next time you go to the beach, bring a camera and take some snapshots of the people you're with at sunset—when they see the results, they'll love how good you made them look.

In low-light settings, including shoots at night, in fog, or on particularly dark and rainy days, manual exposure settings, careful focus, and a high-contrast viewfinder become especially important. Video camcorders can record images in increasingly dark settings, but the images don't always look that good. In very dark environments, some camcorders produce images that look grainy or appear to be made of very small moving spots. If a location is too dark, you may come away with an image that's too grainy to use. A good quality viewfinder can help you avoid this problem. If you see grain in the viewfinder, try to recompose your shot in a brighter area or subtly add lighting to your current location. Large-scale film productions heavily light locations at night, often lighting the facades of all the buildings on a street. If your story necessitates shooting at night, this may be something to consider. I'm not suggesting you bring in floodlights to light up your entire neighborhood, but nighttime outdoor shoots entail specific technical challenges.

---- **N O T E** ----

If you shoot video of a person driving in a car at night, lighting can be a particular challenge. Unless you're in a very bright area, the car may be too dark to get a good exposure. Because the inside of a car offers a limited amount of space, lighting your subject without casting shadows or blinding her can be a challenge (and if you're shooting the driver of a car, blinding her is something to avoid at all costs). One solution filmmakers often employ is to light a driver from below the dashboard. Using a battery-powered light, even a wide-beam flashlight, you can create a light source that makes the shot look like it's being lit by the instrumentation in the car's dash. Positioning the light takes some finesse: putting it in the wrong spot will make your diver look like she's telling scary stories at a campfire. Once you get it right, the results are worth the effort.

In darker shots, depth of field becomes particularly limited (for a detailed discussion of depth of field, see Chapter 2). In bright daylight, a cinematographer can much more easily keep the foreground and background of an image in focus at the same time. At night, the depth of field is more shallow, which may provide you with an effect you like, but also makes it harder to keep all areas of a frame in focus. Automatic focus settings on a camcorder make things even more difficult. If you have objects at different focal lengths—for example, a person closer to the camera lens speaking to someone farther back—the camcorder may not know which person to focus on. In this situation, the camcorder might not focus on the person you want, or even worse, the camcorder's auto focus might shift between the two, changing focus from one to the other at precisely the wrong time. I once worked on a shoot where we interviewed a man sitting in a shady area of his garden. To save money, my boss shot the video himself and left the auto focus on. The shade was dark enough to narrow the depth of field considerably, and the auto focus kept drifting back and forth between the man we were interviewing and the flowers blowing in the breeze around him. It looked like the lens was breathing. This rendered the entire shoot unusable; the changes in focus would have made audiences seasick.

Ultimately, my boss and I used the interview of the man in his garden as voiceover. In the editing room, we paired his recorded audio with a combination of observational footage and still images. On the dry erase board behind us, the editor of a previous project had written:

> "You have to kill your darlings." —Faulkner

> "It's not the pearls that make a necklace; it's the thread." —Flaubert

> "We'll fix it in post." —Anonymous

In this chapter, we covered color casts (daylight is blue, indoor light is yellow) and the ultimate importance of white balance (it calibrates your camera's color balance to the light in your shooting environment). The chapter also discussed aperture settings and why you should always make all the adjustments yourself instead of relying on your camcorder's automatic functions. This chapter also provided tips on shooting outdoors (cloudy days are great) and for lighting indoor shoots (turn off incandescent and fluorescent lights, try using a Chimera and hair lights in interviews).

The next chapter explores shooting for effects. Working with digital video enables you to create complicated special effects that would have been unimaginably expensive only a few years ago. Today, if you have access to a digital camcorder and some editing equipment, you can make anything you want—especially if you take the time to carefully plan your effects sequences before you shoot.

NOTE

Problems with focus and severe over or under exposure are not repairable in postproduction. No matter how good the material is, if it's technically flawed at the most fundamental levels, there's only so much you can do with it. Some white balance problems can be fixed in the color-correction phase, but life is much less complicated (and less expensive) when things go right from the beginning.

Shooting for Effects

5

There's a sequence in *Wayne's World* where Wayne and Garth introduce the audience to a new feature of their show, *chroma key*, and use it to travel to New York, Hawaii, Texas, and Delaware without ever stepping foot outside the studio. Chroma key, the effect they were using, is the same technology that allows meteorologists to stand in front of moving weather maps on the 6 o'clock news.

In *Wayne's World*, the characters stand in front of a chroma key screen and as they pretend to be in New York, they joke about carrying a gun and going to see a Broadway show (which is of course what all New Yorkers do, at least on the weekend).

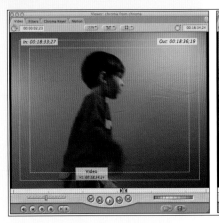  

Figure 5-1. In the first image above left, a boy stands in front of a chroma key screen; the second photo depicts a street scene he will be composited into; the third image shows the boy composited into the background clip.

The images in Figure 5-1 are screen captures from a chroma key sequence created in Final Cut Pro. In the first image, our young star, Luca, is running past a chroma key screen. The center image shows the exterior wide shot into which he will be composited. The final image shows the completed composite: the green background has been removed from the first clip, and the boy has been added to the exterior clip.

Chroma key (also referred to as *blue screen* or *green screen*) involves shooting a person or object in front of a colored screen, then using an effect to remove the colored screen from the shot and replace it with a substitute background. The particular screen colors, *highly saturated* blue or green, are used because they don't show up in nature—regardless of a person's skin tone, you'll never see someone who looks that blue or that green. (For a detailed explanation of what saturated color means, as well as how color functions in digital video, see Chapter 16.) In postproduction, a technician identifies the color of the blue or green screen in a shot, and then removes that color from the footage. This process is called *keying*. When you key out, or remove, a color using a non-linear editing system, all the pixels containing that color become transparent. (Often, multiple filters have to be applied in the postproduction process to avoid an annoying glow or shadow around the object that makes the background look fake. For more detail, see Chapter 12.)

In the next step, a technician adds a still image or a video clip to a lower track in the editing program's timeline, creating a new background for the shot. Because the keyed clip now contains areas that are transparent, any still or moving image the technician adds as a background clip will show through. The process of combining two (or more) video images into one is called *compositing*. If you've worked with still-image editing programs, such as Adobe Photoshop, you may already be familiar with compositing and transparency.

The compositing features of digital video programs, such as Final Cut Pro or After Effects, present tremendous opportunities for independent filmmak-

> **NOTE**
>
> *Chroma key is not only used in older films and television shows. Many recent blockbusters, including* Spiderman *and* Lord of the Rings, *make extensive use of chroma key to create eye-popping effects and stunning visual environments. This has created an entirely new form of filmmaking in which actors interact with imaginary costars who have yet to be composited into the frame, and directors create entirely new worlds out of nothing more than computer-generated images, or CGI. A hefty portion of* Sky Captain and the World of Tomorrow *was done this way.*

ers. You can composite still or moving images into any shot, using a variety of techniques. Once you master the process, images will blend together so smoothly that audiences will never know they're looking at a composite. Wayne and Garth would be proud.

Preparing a Chroma Key Shoot

In the recent past, working with blue screen or green screen required considerable resources, for example, finding an available screen. If you knew the people at your local public access station, you might have been able to use theirs. If not, or if the screen was already booked by someone else, you would have to spend the money to rent time at a production facility. Now, prosumer technology makes it affordable to set up a good quality screen in your home studio or even on location. Many film and video equipment stores sell chroma key screens at accessible prices. A company called Lastolite sells collapsible chroma key backgrounds that it advertises as "completely crease free." Creased backgrounds can be a problem for chroma key effects because the creases produce shadows, making some areas darker and others lighter—this results in a multicolored background instead of a solid color background (solid backgrounds are easier to key out, because they contain only one color to identify and remove). Working with a creased background isn't the end of the world, but avoiding creases might save you a few gray hairs.

To create the chroma key effects for this book, I used a 5×7 foot background, from a Colombian company called Botero, that folds into a 26-inch disc for travel and storage. Some collapsible backgrounds are also reversible, blue on one side and green on the other, which lets you switch colors depending on your actor's wardrobe. (As of this writing, a single-color Botero chroma key background costs $63.50; a reversible model, such as that in Figure 5-2, costs $79.95.) If you shoot a man wearing a blue suit in front of a blue screen, parts of his suit may become transparent when you key out the screen (you may have noticed that this happens on television sometimes when a new person does the weather forecast and doesn't quite know what to wear).

Various companies also sell reversible chroma key curtains as well as rolls of chroma key "seamless" paper backgrounds. ("Seamless" means that the material curves onto the floor, without making a distinct angle

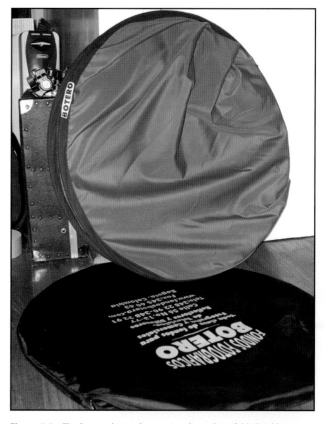

Figure 5-2. The Botero chroma key screen, shown here folded, is blue on one side and green on the other. The same company sells slightly less expensive versions in either green or blue.

that would produce shadows.) Other companies, and some hobby stores, also sell large pieces of blue or green cloth you can use as a chroma key background at a much lower price; just be sure to stretch the material tight enough to avoid wrinkles and creases. If you're looking to create a large and permanent chroma key stage, you can use chroma key paint. Rosco International sells specially formulated chroma key paint for $44.95 a gallon. According to the company, a gallon of paint covers 300 square feet. A dedicated chroma key stage can be helpful if you want to record large areas of action, such as a person running across the frame, or if you want to record a number of people in the frame together.

Lighting your chroma key shoot

Just as a crease-free background makes chroma key effects easier to execute, so does evenly lighting your background. Even if you work with the perfect solid-color background, shadows or inadequate lighting can make key effects more difficult, if not impossible.

The background

An editing system will likely interpret lighter and darker areas as multiple colors, creating more than one color for you to key out. If an editing system interprets some of the darker areas of your background as similar in color to darker areas of your actor's hair or wardrobe, the editing system will key out pieces of your actor along with the chroma key background you're trying to remove. This can be really frustrating unless you're doing it on purpose, but fortunately, there are ways around the problem.

If you have Final Cut Pro installed on a laptop, the software's documentation suggests bringing your laptop to the shoot, connecting your camera via the FireWire port, and then using Final Cut's *waveform monitor* and *vectorscopes* to ensure your background is evenly lit (you can perform similar tests using Premiere Pro, Avid XPress DV, and Sony Vegas). Waveform monitors and vectorscopes (addressed in detail in Chapter 16) measure the color and brightness information in a video signal, and will reliably point out any potentially problematic areas in your shot. You can use the software to measure color information with, and without, your actor standing in front of a chroma key background. As they say, an ounce of prevention is worth a pound of cure. Even if it takes considerable time and effort to fine-tune your lighting before you start to shoot, it can save you immeasurable frustration in the long run if it makes your effect easier to execute (not to mention the money you save by not having to reshoot the scene to correct a problem). Like everything else, the process will go more quickly as you get more practice.

To light the chroma key background in Figure 5-3, we placed very powerful 600-watt tungsten lights at either end and positioned them to light the background as evenly as possible. As you can see, the light is not perfectly

NOTE

If you're planning a chroma key effect to composite a person into an outdoor scene, shooting the outdoor footage on a cloudy day can make lighting much easier. As described in the last chapter, cloudy weather produces an even light without shadows. It's much easier to match the lighting in a shadow-free clip, because you don't have to worry as much about the direction of the light source.

We purposely waited for a cloudy day to shoot the street scene exteriors for this book (we shot in New York in December, when cloudy days are plentiful—if you're working in a sunnier climate you may not have this option). In addition to the issue of shadows, the color temperature of an outdoor scene varies by time of day and is especially noticeable in shots that contain bright sunlight. If the color cast of the actor doesn't match the color cast of the background clip, the audience will notice. Cloudy days create an even, flat light that's much easier to reproduce in front of a chroma key screen.

even. Some areas are slightly brighter than others, but there are no problematic areas of extreme darkness or shadow.

The actor

Once you have your background evenly lit, it's time to light your actor, preferably without casting shadows on your carefully lit background. One preventative measure you can take is positioning your actor at a distance from the background, as in Figure 5-4. This not only helps prevent shadows, but it also helps to prevent the color of your screen from reflecting, or *spilling*, onto your actor. If blue or green light reflects off your background onto a person in the scene, you may inadvertently key out parts of his body along with the background during the key process.

It's important to try and match the lighting of the clip you plan to composite your actor into. If you're compositing a person into a shot containing a weather map, or another type of animated background that isn't designed to look realistic, matching the lighting scheme won't be too much of an issue. If, however, you're attempting to create a realistic looking composite, the lighting on your actor needs to match the lighting in your background video clip. If you plan on compositing a person into footage that contains any type of shadows, it's important to keep the direction and the darkness of those shadows in mind. If you have noticeably flat lighting on a person's face, and dark shadows in the clip you're compositing that person into (or vice versa), the shadows won't match. As a result, the composite won't look believable to the audience, even if your chroma key is otherwise technically flawless. You can work with color correction tools to compensate for differences in color temperature or overall brightness (again, these are addressed in Chapter 16), but shadows and the direction of a light source are not elements you can readily change using prosumer video-editing systems.

Wardrobe, hair, and smoke

Cleaner edges yield cleaner key effects. As a result, a person with neatly arranged hair is much easier to key out of a background than a person with wild wisps of hair going off in every direction—the fine strands can result in a fuzzy edge that's harder for a computer to separate from a chroma

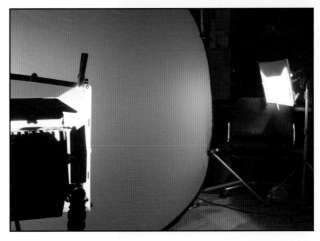

Figure 5-3. The trick to chroma key is good lighting. If a green or blue screen background isn't lit brightly enough, it becomes much harder to key out.

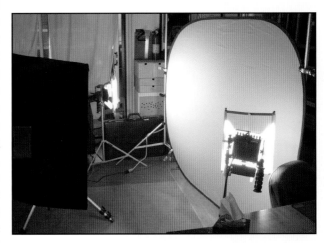

Figure 5-4. To light our actor, we placed a Chimera (it's the large light in the left of this photo) as far from the background as possible. Our goal was to evenly light Luca as he ran past the screen, without casting shadows on the background.

key background. (If the computer cannot effectively separate a foreground object from the background chroma key screen, it leaves a thin halo of color around the edges.) Shaggy wardrobe materials, like a fuzzy sweater, a fleecy jacket, or clothing with fringes can be especially hard to work with as well. Some types of dangling jewelry, such as fine strands of gold in a necklace or earrings, can also be problematic if they hang away from your actor's body.

Clean edges are particularly important when you're working with DV. Prosumer mini DV is a lightly compressed format—the compression takes place in the camera, and most of the time audiences and filmmakers never notice. The compression, however, makes it even harder to separate fine or fuzzy edges from a background using a chroma key effect. If your story absolutely requires hard-to-key images, such the hair or wardrobe elements just described, or if it contains translucent visuals like smoke, a format other than DV might make a better choice.

Uncompressed standard-definition formats like Betacam SP or more lightly compressed high-definition formats allow more latitude in key effects. Don't worry though: if you're careful about hair, clothing, and lighting you can still get good results with DV. If you're not sure, do some tests. Shoot some footage of your actor in a dress rehearsal, and try out the composite techniques you'd like to use. With a little experimentation, forethought, and some effort, you'll get the results you're looking for.

Framing Images with Composites in Mind

Just as the shadows and lighting in your key shot need to match the shadows and lighting in your background, the content and framing of your foreground clip needs to work well enough with the background clip that the combination is believable. An obvious example is if you plan to composite someone into a shot on the beach, they shouldn't be wearing a winter coat. (Unless you want to do this for effect; for example, to imply humor, irony, or someone who is crazy or doesn't plan well.) There are also more subtle considerations, including camera angle, framing (wide shot, close up, etc.), and the placement of elements within the frame.

If you shoot an actor against a blue or green screen and then key out the background, it's often easier to composite her into a shot if you don't show her entire body from head to toe (Figure 5-5). Have you ever noticed that you almost never see the feet of a meteorologist when he's standing in front of an animated weather map? Part of the reason is there's no real reason to show a person's shoes as he forecasts the weather, but it would also make the composite more difficult. The shot would require not only placing someone in front of a green or blue screen background, but placing him on a blue or green floor as well, and then lighting everything without producing shadows—no easy task.

> **NOTE**
>
> *If you find that your original framing causes you difficulty when it comes time to edit your composite sequence together, you can use the scale, mask, and reposition techniques described in Chapter 12. Chapter 12 also shows you step by step how to create very slick composite effects, including chroma key.*

Figure 5-5. We shot our actor so that his feet are never shown in the frame.

There are also aesthetic considerations to not framing head-to-toe chroma key shots. If the bottom of the newscaster's body is cut off by the frame of the screen, the viewer intuitively understands that the person is standing in front of a weather map and that part of her body has been *framed out* of the shot. If we see the newscaster's feet, and the floor and background are keyed out, she would look like she was floating in space. This would either look like a mistake, or would just plain look strange. (Again, there might be a reason you'd want to do this. Once you learn these rules, you can break them.)

Likewise, if you're compositing an actor into a landscape or street scene, it becomes harder to composite the actor in a way that looks plausible if you show his entire body. A good solution is to compose a more close-up shot, perhaps showing him from the waist or chest up. That way, the actor looks like he's in the foreground of your frame, with a realistic looking background behind him. Viewers automatically make allowances for subtle visual differences in foreground and background images, and are less likely to notice small inconsistencies in lighting or color temperature—if a woman is standing across the street from a building, it makes sense that the shadows might fall at a slightly different angle on her face than on the structure behind her. If the person is supposed to be right next to an element in your composite background, slight inconsistencies become a problem.

Chroma key effects are not the only composite technique at your disposal when you work with DV. You can also add a variety of still images to the foreground or background of your compositions, including photographs and titles. As you develop more complicated composites, it helps to frame your shots with your finished image in mind. For example, if you plan on creating titles in a still image creation program such as Adobe Photoshop or Adobe Illustrator, or if you plan on adding special effects to your titles using a more advanced program like After Effects or Motion, you might decide to frame a shot and leave a space for the title. HBO's *Six Feet Under* brilliantly incorporates titles into the frame—some appear distorted as if you were viewing them through glass, others rise and fall in time with the parts of embalming apparatus. Framing a shot with space for the title text enables the show's design staff to integrate the titles into the opening sequence and create a really slick, coordinated effect. (We explore advanced title-creation techniques in Chapters 14 and 15.)

Planning a Matte Effect

As described earlier, a chroma key effect removes areas of color from the background of a shot, enabling you to substitute another image in its place. A *matte effect* defines an area of the screen, in the shape of your choice, that becomes transparent and enables another video image to show through. The shape that defines the transparent area is called a *matte*. Final Cut Pro

> **NOTE**
>
> *HBO has a great series of promotional spots in which characters from different shows are composited together in conversation. In one clip, Larry David discusses intimate details of his sexual dysfunction with Adriana La Cerva of* The Sopranos. *In another, a bathrobe clad Tony Soprano walks in for breakfast with the Fisher family of* Six Feet Under. *All of the pairings are extremely witty and skillfully executed, but visually some of the composites simply work better than others. The reasons are simple: lighting and framing. The composites are done by experienced professionals working with sophisticated equipment, and from a technical standpoint, they're all brilliant. But if the lighting in one clip doesn't match the other, there's only so much an editor can do. Similarly with framing, not all images look like they were shot at the same time. Don't get me wrong, I could watch these HBO spots all day, but when you start planning a composite sequence, think about all the details it takes to bring your work to the level of quality you're looking for.*

NOTE

I was sufficiently inspired by seeing the original Godzilla that I created the demonstration sequence in Chapter 12. In that sequence, a giant two-year-old Luca (pictured throughout this chapter) looms over New York's historic Lower East Side. Like the matte in the original Godzilla, I used the tops of the buildings to delineate the edge of one shot and the beginning of another. As a result, traffic flows freely through the bottom half of the frame while gigantic Luca rampages across the top. The buildings naturally fill the bottom half of the frame, and it makes sense visually that an oversized toddler would appear behind them. As in the earlier example of framing out a meteorologist's feet, the buildings also provide a natural way to obscure the lower portion of Luca's body, which helps make a more natural looking composite. To give credit where credit is due, the toddler rampage sequence in this book was partially inspired by another two-year-old, who ran across our picnic blanket last Fourth of July. He stormed straight through the center of our meal, laying waste to paper plates, plastic flatware, and marinated chicken kebobs. When his father, who was dining with us, finally caught up with him, he turned to me and said, "He's just like Godzilla isn't he?"

and Premiere both ship with ready made matte effects you can apply as filters to create round, square, or diamond-shaped mattes. Both programs also enable you to create an (unfortunately named) *garbage matte* to create custom matte shapes using 4 or 8 points on the screen. Last but not certainly not least, you can use a *travel matte* to create a matte effect in the exact shape of a complex graphic you've produced in a program such as Photoshop or Illustrator.

Mattes have been used in films for years, and particularly good examples can be found in the *Godzilla* movies of the 1950s. One afternoon, I took a break from writing this book and went to my favorite independent theater, Film Forum on West Houston Street in Manhattan, to see a reissued print of the original *Godzilla* (not the U.S. version with Raymond Burr edited into an otherwise all Japanese cast). I sat behind two grade school–age kids, who to my surprise, loved the movie and even remarked about how "real" everything looked. During one of the climactic scenes, a man in a Godzilla suit rampages through Tokyo (this was the part the kids liked best). As I watched the scene, I realized it was a very well-constructed matte. The bottom half of the screen showed an actual Tokyo cityscape complete with traffic, firefighters, and soldiers fleeing the giant monster. The top half of the screen showed Godzilla breathing fire. The film's effects designers created the composite by cutting a matte in the shape of the tops of the buildings in the cityscape (the original was released in 1954; in those days, people cut physical mattes to divide the frame into multiple shots). Because the effect displayed two separately filmed actions, the filmmakers could show actual motion-picture footage of the city in one part of the screen and artfully add action footage of Godzilla in another. The end result is a composite that holds up to discerning audiences more than 50 years after it was first released—and who do you know that's more critical of film effects than an 11-year-old?

Composing reflections

In Roman Polanski's *Chinatown*, there's a brilliant shot of Jack Nicholson watching a bad guy through a pair of binoculars. As Nicholson holds the binoculars to his eyes, the audience sees the bad guy's reflection in both lenses. Reflections are a great tool for filmmakers—whether in mirrors, car windshields, eye glasses, or Nicholson's binoculars. They allow a director to reveal action to the audience in the most artful of ways. By watching the reflection in the binoculars, audience members feel like they, too, are spying on the villain, which heightens the drama. (In retaliation for being "nosy," the bad guy later cuts off part of Nicholson's nose.) Thanks to a digital editing system's compositing techniques, these kinds of effects are becoming increasingly accessible to independents making small-budget productions.

As always, careful planning beforehand goes a long way. Before you go out to shoot, think about what shots will work together as composites. Later, once you're in production, think about how you can frame shots so they'll work as composite elements.

For example, if you want to show a person's reflection in a pair of eyeglasses:

1. Plan and compose a good close-up shot to use as your reflection.

2. Record the close-up as large as you can (don't try to record the shot at a size small enough to fit into a pair of eyeglass lenses).

3. Shrink the shot down to eyeglass size using the scaling tools in Final Cut, Premiere, or After Effects (working at full size and then shrinking an image provides you with much more detail and control than attempting to shoot something really small).

4. To make the reflections look more realistic, and less like something you pasted in, try different composite modes (composite modes, which are described in Chapter 11, determine how elements blend together when composited into a shot).

5. Composite the same reflection clip into both lenses—that way, if the reflection contains movement, the action will be the same on the left and the right side.

Working with a tripod

The compositing capabilities of today's prosumer digital editing programs provide you with a tremendous amount of control over transparency, placement, and even motion. Working with a tripod allows you to shoot stable images and take full advantage of the control these digital systems afford you as a director.

It almost seems ironic, but even though you have all this computer power at your fingertips, something as low-tech as a tripod can make or break your shot. If you plan to add a matte background, using a tripod will make your life significantly easier. If you shoot a street scene and plan to composite another image behind the buildings in your shot, the process becomes much simpler when the camera is held steady. Regardless of how much motion is in the street, the composite becomes infinitely easier if the buildings in the background don't move.

The next chapter introduces techniques to keep your camera steady while executing complicated camera movements. Like so many other things in this book, the technology behind the Steadicam (which enables filmmakers to stabilize a camera during even the most hectic action shots) is now available to you in a prosumer version. The next chapter shows you how to use it.

— NOTE —

Camera angle is another important consideration in creating reflections. If you plan on compositing a reflection into a car windshield, film the windshield from an angle that makes the reflection believable. Rather than simply filming the car's windshield straight on, you might do better shooting from the side, so you can simulate objects passing by to create a feeling of movement.

Working with Specialized Camera Mounts

6

Having watched years of professionally produced movies and television shows, audiences have high expectations for what they see on screen—they take for granted that all camera movements will be smooth and every shot will be steady. If your camera has the slightest shake, people will notice. In addition to mastering the technical aspects of digital video (frame rate, aspect ratio, color balance), producing a good show requires precise camera control to get the motion effects you want.

For years, this meant using expensive rigs and elaborate setups ranging from specially equipped trucks, to cameras mounted on dollies that rolled along a track, to robotic cranes that moved cameras by remote control. Needless to say, these were not cheap. The affordable options were to shoot handheld, which resulted in shaky images, or to mount the camera on a stationary tripod, which produced smooth pans and tilts but limited mobility. Then came the Steadicam.

Stabilizing a Moving Camera

A Steadicam (*www.steadicam.com*), shown in Figure 6-1, is a camera stabilization system that mounts a camera on a rig, called a sled, and balances the camera using a monitor and battery pack that hang underneath and serve as counterweights. The rig lowers the camera's center of gravity, and the design of the Steadicam isolates the camera itself from the operator's movement.

In the full, professional versions, the Steadicam attaches to a harness, which distributes the camera's weight and allows a cinematogra-

Figure 6-1. A Steadicam and its operator.

pher to operate a large, heavy camera for extended periods of time. (Even with the harness, Steadicam operators have to be fairly physically strong.) The Steadicam sled connects to the harness through a spring-loaded arm that enables an operator to move freely without causing any *camera bobbles*, or shaky images (a practiced operator can even run or climb stairs). The resulting shots look as if they were captured by a camera floating in midair. A talented cinematographer can achieve fairly stable results with a handheld camera, but in close-up shots, even the slightest camera movement becomes especially noticeable.

A high-end Steadicam is by no means cheap; the list price for the top model is more than $65,000, but with the advent of high-quality prosumer DV camcorders, the company began to offer more affordable models. The scaled-down Steadicam Mini, for cameras that weigh 5 to 15 lbs (such as the 6.5 lb Cannon XL2) lists for $5,500. There's also a handheld Steadicam JR for cameras 4 lbs or less (such as the Sony VX2100 or the Panasonic AGDVX100A). The Steadicam JR, which does not attach to a harness, lists for well under $1,000, falling solidly into the prosumer price range.

The Steadicam JR's setup manual (available online free of charge at *http://www.steadicam.com*) compares the function of a Steadicam to balancing a cereal bowl on the tip of your finger, as in Figure 6-2. The manual explains that balancing the bowl is difficult, because its center of gravity is higher than your finger. If you turn the bowl upside down, the manual points out, balancing the bowl becomes significantly easier (even while moving your hand) because the bowl's center of gravity is lower than the tip of your finger. Even if you have no interest in balancing a bowl on your finger, the comparison provides a great explanation of the principles behind the Steadicam's operation. Placing the weight of a monitor and battery pack at the bottom of the rig helps balance the weight of the camera on top, making it easier to stabilize.

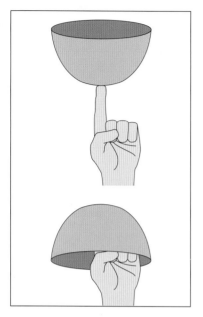

Figure 6-2. A person attempting to balance a bowl on the tip of a finger. In the first image, the bowl is upright, and the person's finger is on the bottom of the bowl on the outside. In the second image, the bowl has been turned upside down, and the person's finger is inside the bowl, which lowers the bowl's center of gravity and makes it easier to balance.

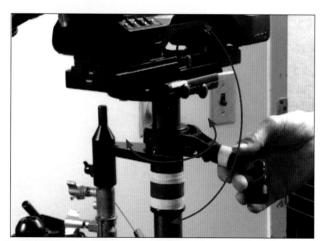

Figure 6-3. The gimble is the "guts" of the Steadicam.

In both the handheld and body-mounted versions, the sled connects to the rest of the rig through a gimble, which provides a full, flexible range of movement. The construction of the gimble, which Steadicam operator Lou Rosenberg describes as the "guts" of the Steadicam, enables an operator to smoothly tilt a camera forward and backward, lean it from side to side, pan the camera as if it were on a tripod, or execute any of these moves in combination. A gimble is shown in Figure 6-3.

Steadicam is a brand name that, in everyday speech, has erroneously become a generic word for any type of device that keeps a camera stable during complex movements—much like Kleenex has come to mean tissue. (Some companies love this. It means they are at the top of their field. Others hate it and feel it dilutes their trademark.) Other companies offer competing stabilization systems, both hand-held and harness based. For example, a company called Glidecam sells the Glidecam 2000 Pro handheld stabilizer for cameras that weigh up to 6 lbs, and the handheld Glidecam 4000 Pro for cameras that weigh 4 to 10 lbs. These two systems retail for $369 and $499, respectively. The same company also sells the Glidecam V-8, a harness mounted system for cameras that weigh up to 10 lbs, which lists for less than $3,000. Other companies, such as VariZoom and Hollywood Lite, also offer similar products.

The expanded vocabulary of Steadicam shots

The advent of the Steadicam introduced a new kind of shooting. In love scenes and music videos, it's become common to see the camera make smooth 360 degree revolutions around the star, providing a dizzyingly beautiful shot that never would have been possible without a stabilization system that was easy to move. In the pre-Steadicam era, filmmakers could circle around a subject by mounting the camera on a dolly, and placing the dolly on a circular track. Track-mounted dollies can move in extremely smooth circles, so a director using one could get great shots, but the circular movement was fixed—the camera could only move around the set path of the track, and if the actor moved beyond the circumference of the track the shot wouldn't work. If a director wanted to stage shots in more than one location, or from multiple angles, the crew would have to reassemble the track for each new shot. As you can imagine, this limited the type of shots available to a director, even in the most carefully staged production, and made the use of tracking shots generally impractical for most documentary work.

The Steadicam and Glidecam, however, enabled smooth movement through crowds; stable camera work on uneven surfaces like stairs or outdoors on rough terrain; and created a whole new type of chase scene where the Steadicam operator runs along side the actors without shaking the camera. The Steadicam has also enabled filmmakers to shoot great tracking shots from any type of car, pickup truck, or motorcycle, even from a golf cart. Large-budget shoots often employ cranes or cherry pickers for aerial shots, and artful cinematographers can combine them with a Steadicam to get truly impressive results—more than one episode of *ER* contains a breath-taking shot captured by a Steadicam operator who walks onto or steps off of a moving crane partway through a take.

Affordable camera stabilization has also made its mark on small-budget independent works, student films, and even weddings and birthday party videos. Fluid, stable shots have become the norm to such an extent that jerky camera movement is often reserved for especially chaotic chase scenes or for building tension in horror films. In *The Shining*, director Stanley Kubrick followed his young protagonist with a slightly wobbly camera as he rode laps around the empty hotel on his low-rider tricycle, evoking remarkably

---NOTE---

There is a web site called "$14 Steadycam [sic]—The Poor Man's Steadicam" (http://www-2.cs.cmu.edu/~johnny/steadycam/) that shows you how to build your own camera stabilization device. The site gets its name because the parts, most of which are available at your local hardware store, cost about $14. While this isn't a true Steadicam ("Steadicam" is a trademark of the The Tiffen Company, LLC), this "$14 Steadycam" offers a possible low-budget solution, especially if you like to build things yourself.

SIDEBAR

Truck Versus Dolly

Just as there are names for different framings (for example, MS, CU, ECU, and others described in Chapter 3), there are names for different camera movements. Many of the names come from the equipment traditionally used to execute a specific type of movement, such as a truck or a dolly. With the advent of stabilization devices like the Steadicam, trucking shots and dolly shots can now be made without an actual truck or a dolly, but the names still hold.

ask a Steadicam operator to truck from left to right across a room.

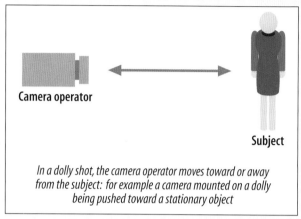

In a dolly shot, the camera operator moves toward or away from the subject: for example a camera mounted on a dolly being pushed toward a stationary object

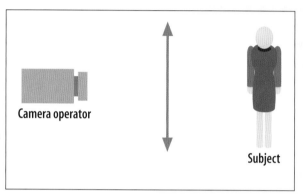

A trucking shot simulates the movement of a camera mounted on a truck, and moves the camera parallel to the action. Trucking shots don't necessarily move the camera closer to or farther away from a subject. A director might

A dolly shot moves a camera closer to or farther away from the subject. As you can guess from the name, it simulates the movement of a camera mounted on a dolly. To "dolly in," is to move toward something, to "dolly out," is to move away. (See "Using a Camera Dolly" later in this chapter.)

Dolly and truck shots can also be used in combination with pans and tilts. A pan moves the lens of a camera from side to side; a tilt moves the lens up or down.

unsettling feelings in his audience long before anything overtly frightening showed up onscreen. Stable camera work has become such a convention that the Coen brothers invented a Shaky-cam, a camera mounted on a plank of wood with a person holding each end, designed to produce noticeably un-Steadicam like images for dramatic effect.

Additional Steadicam and Glidecam Mounts

If you're not satisfied with leaning a stabilized camcorder outside a car window to get a shot, Glidecam offers a vehicle mount you can use to affix your Glidecam V-8 to the outside of a truck or another moving vehicle—it attaches the Glidecam to a fixed, vehicle-mounted post instead of a harness worn by a cinematographer. This is an affordable way to get an actual trucking shot.

If you want to simply attach a camera to the outside of a car, you can buy something called a "Sticky Pod" for slightly more than $100, which affixes a camera to the outside of a vehicle using suction cups (yes, suction cups—I don't know anyone who's used one so I'm not endorsing it, I'm just fascinated by the idea). The Sticky Pod mounts a camera on a post (Figure 6-4), which is attached to a small platform with suction cups at the corners. The company advertises that its product "will stick to any car, truck, boat, plane, or motorcycle," and stay put at speeds of more than 110 miles per hour. If you're worried about losing an expensive camera at high speeds, which is not an unreasonable fear, you can mount a camera inside a car facing your actors or the road. (You could also try using an older or cheaper camera and accepting the risk—you might get some great tape, if it doesn't cost you your camera.)

Mounting a camera inside or outside a car enables you to take point-of-view shots of the vehicle moving through traffic, or of the driver and passengers through the car's windows. Shooting from outside through the windshield of a moving car enables you to capture great reflections that add to the realism of a shot. Of course, you also have the option of filming a stationary car and compositing in reflections or dramatic chroma key backgrounds (see Chapters 5 and 12). Directors often combine Steadicam footage with chroma key effects to create driving sequences; this provides more control to the filmmaker and helps safeguard against high-speed crashes. The effects don't have to look fake, especially if you're careful about lighting and other considerations mentioned in Chapter 5. Much of *The Fast and the Furious* was shot using composite techniques to avoid placing the cast and crew in overly dangerous situations, and it still looks really cool.

Both Steadicam and Glidecam offer "low mount" attachments, to help you shoot stable footage at low angles. These can be especially helpful for shooting ground-level shots, for example, filming small animals at their eye level, or shooting low-angle shots of your actors—shots of footsteps moving toward or away from the camera are staples in suspense films. If your shoot calls for stationary low shots you can also use a short tripod, often referred to as a set of "baby legs." Baby legs offer the same benefit as a tripod, but they're much shorter, so they enable you to stabilize a camera only a few inches off the ground, as in Figure 6-5. (They also cost considerably less than a Steadicam or Glidecam.) If you don't have baby legs available, you can stabilize a camera by nestling it in some sandbags, or even, in a pinch, wedging it between some coats or jackets. In the days before affordable Steadicam-type stabilizers, I made a student film that opened with a shot from a car driving down a palm tree–lined road. A classmate filmed the shot by leaning out through the car's sunroof holding a 16 mm film camera against a sandbag. Very low tech, and rather uncomfortable for my classmate, but the shot turned out great.

Glidecam also sells a portable camera crane, the Camcrane 200 (Figure 6-6), that mounts on a tripod and allows you to raise and lower your camera while moving it in 360-degree arcs. It retails for less than $550. Camera cranes extend the range of your cinematographer by placing a camera at the end of an adjustable boom. This enables filmmakers to move a camera into

The POV Shot

The term POV means a shot taken from the point of view of a character in a film. Rather than simply displaying a character on a screen, a filmmaker can use POV shots to draw the audience further into the action. By showing the world through the eyes of someone in the film, a POV shot can add layers of complexity, detail, and power to a sequence. (Spike Jonze uses a great POV shot at the end of *Being John Malkovich*.)

A Steadicam is especially helpful for creating POV shots because it provides a camera operator with tremendous flexibility and control. A good Steadicam operator can reproduce a full range of actors' movements (including walking into a room, climbing stairs, and even chasing or running away from someone) that can help audiences feel like they're part of the story.

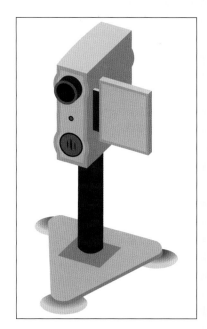

Figure 6-4. With the aid of suction cups, a Sticky Pod attaches a camera to a vehicle.

Figure 6-5. "Baby legs," essentially a short tripod.

Figure 6-6. The Camcrane 200.

parts of a scene where a camera operator might get in the way, or to simply film from an angle beyond a cinematographer's physical reach. A company called Porta-Jib (*www.porta-jib.com*) makes a more sophisticated (and more expensive) jib arm-type crane that enables cinematographers to extend and retract the crane's arm during a shot, dramatically expanding the possible options. The Porta-Jib is priced at the higher end of prosumer products (it lists for more than $2,000) but might be worth exploring if you feel it would really help your film.

Both the Camcrane and Porta-Jib can be used in stationary locations or on camera dollies that accommodate a tripod. Losmandy, the same company that makes the Porta-Jib, also sells a Spider Dolly (Figure 6-7), which is essentially a base that accommodates a tripod for tracking shots. Losmandy sells both three- and four-wheeled versions of the Spider Dolly. Either can be used on a track or with floor wheels (similar to the wheels on a supermarket shopping cart but designed to move much more smoothly). As you may have guessed, the company also sells track. Their FlexTrack comes in 40-foot spools that filmmakers can position in straight or curved paths. According to the company, a looped 40-foot section of track provides a 17-foot run. You can also place two 40-foot pieces end to end for really long

SIDEBAR

Finding an Equipment Rental House in Your Area

Not all equipment rental facilities are worth your business. Some may rent out older or even damaged equipment, some may not want to work with an independent on a small budget. Ask around before you rent, and don't waste your time or money on a business if someone you know had a bad experience with them.

Here are some tips to aid you in finding a rental house:

- Join a professional association of filmmakers. Some rent equipment to members at discounted prices, and most help you get discounts on rentals, tape stock, and professional services from other businesses in your area. The Film Arts Foundation (*www.filmarts.org*) and Bay Area Video Coalition (*www.bavc.org*) in San Francisco, Association of Independent Video and Filmmakers (*www.aivf.org*) in New York, and International Documentary Association in Los Angeles (*www. documentary.org*) are all large organizations open to membership from people around the world.

- Contact a local arts organization and ask if they have any recommendations. There are local organizations from the biggest cities to small towns across the country. Chances are somebody at the local office will have some good suggestions, or at least be able to point you in the right direction.

- Ask your friends. The people you know are always your greatest resource. If you have friends who've rented equipment, ask who they rented from and what their experience was like—it's the best way to get a straight answer.

tracking shots. A three-leg Spider Dolly lists for about $1,000 (the dolly with floor wheels costs slightly more than the track version). The four-wheel dolly lists for more than $3,000 and comes with a rideable platform and swiveling seat, so your camera operator can ride along with the camera and turn as needed.

Just about any piece of filmmaking equipment can be rented by the day or the week (see "Finding an Equipment Rental House in Your Area"). Even if you live in a fairly small city, there's more than likely a rental house nearby (if you live in commuting distance of New York or Los Angeles you may have more rental options than you know what to do with). Depending on your circumstances, renting equipment can be far more cost effective than purchasing. For example, if you only need to use a $3,000 dolly for one day, it doesn't make too much sense to buy it. Even if you know you'll be using an item often enough to justify a purchase, you might still want to rent one and use it in the field before you commit to a particular model or brand.

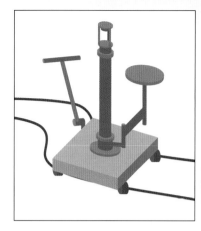

Figure 6-7. A Losmandy Spider Dolly and track.

Using a Camera Dolly

The arrival of the Steadicam on movie sets has by no means made the camera dolly obsolete. Some of independent film's most creative directors continue to use dolly shots to achieve specific effects.

There's a key scene in Spike Lee's *25th Hour*, when a 17-year-old high school student played by Anna Paquin glides dreamily through a nightclub on her way to seducing her awkwardly bookish English teacher, played by Philip Seymour Hoffman. Paquin is standing on the same dolly as the camera, which frames her from the chest up, so she maintains a fixed distance from the camera as the furniture and the people in the club move around her. She seems to float. As she moves, the ambient light on her face changes with her surroundings, highlighting the surreal motion—if she were walking, the camera wouldn't remain at her perfect eye level through the entire shot, and the distance between her and the camera would change slightly as she took each step. Lee is a big fan of this kind of shot (he includes one in just about all of his films) and uses the actor-on-dolly technique again as the scene winds down. The next time, Hoffman stands on the dolly staring up at the camera, looking seasick and bemused as the camera moves with him away from the bathroom where he has just clumsily kissed his student. As he glides through space with the camera, Hoffman stands still, holding the same facial expression and not moving his body at all. The shot produces a similarly surreal effect as Paquin's earlier dolly appearance but with much darker emotional overtones.

Martin Scorsese uses similar dolly techniques in his films, careening through a bar with a drunken Robert DeNiro in *Mean Streets*, and using a point-of-view dolly shot in *Goodfellas* to introduce the audience to the extended family of mobsters as Ray Liotta narrates. Lee and Scorsese use these techniques to draw the audience in and make people feel like they're part of the action they're watching onscreen. In *Goodfellas*, the mobsters all look slightly off camera as they say hello to Liotta's character, Henry Hill. The audience sees the action, and viewers feel as if they were walking through the bar along side Liotta, living the story firsthand.

Using shots like these doesn't in itself make you the next Spike Lee or Martin Scorsese (their work also contains a fair amount of cinematic genius, not to mention tight writing, and stellar actors), but thinking about camera work is a great way to define the visual feel of your film.

Lee and Scorsese are also both masters of using audio to tell a story. Their films make exceptional use of sound as a tool to move a story forward and engage an audience. You can do the same thing. The next chapter of this book explores the history of audio recording in film and video and introduces various audio production concepts available to you as a filmmaker. Chapter 8 examines ways you can use audio equipment to make sure your film sounds as good as you want it to.

SIDEBAR

Negotiating a Price

All rental houses have a card rate, which they generally post on their web site. This is the advertised rate at which they rent out equipment, but often it's just a starting point for discussion. The following tips will help you get the most for your money:

- Look up the card rate for each rental house before you call, so you know their advertised price before you talk to a rental agent.

- Always ask for a lower price. Rental agents get paid on commission, and they all know they're better off renting something to you at a lower price than letting it sit on a shelf. If you ask nicely and become a repeat customer of the same agent, he may even offer you bigger discounts in the future to keep your business.

- If you rent on a Friday, rental houses often let you keep the equipment over the weekend and only charge you for a one-day rental. Shooting on the weekends can bring your rental costs way down.

- Rental houses generally work on a four-day week, meaning that if you rent for a full week (Monday through Friday), they only charge you for four days instead of five. If you rent for a full week, a good rental house will also let you keep the equipment over the weekend at no additional charge. If you can consolidate your schedule and shoot over several consecutive days in a one-week period, instead of scheduling isolated one-day shoots, you can save yourself considerable amounts of money.

- Don't pay for days when you're not shooting. If you have an early morning shoot, a good rental house will let you pick up the equipment the day before and charge you only for the day you shoot.

- Check all the equipment at the rental house when you pick it up and be sure it works before you leave the building. Even a good rental house will sometimes make a mistake and will gladly fix an order that isn't right. Also, if a piece of equipment is damaged or something is missing from your order, let the staff know before you leave—if you don't, you have no way to prove you're not at fault.

- Bring the equipment back on time (both to avoid late charges and to establish yourself as a good customer) and always get a receipt.

Recording Audio, an Overview

7

Film doesn't record sound. Just like audio tape doesn't capture images, 35 mm and 16 mm film simply don't capture any audio. Unlike working in DV, which captures images and sound together, directors shooting film have always captured picture and audio separately, using multiple pieces of equipment. Although digital video now enables filmmakers to record images and sound to the same tape, many filmmakers still use specialized audio-recording equipment, because it provides greater control and flexibility.

In addition, directors often record a scene, and then after viewing the footage, decide to record additional audio to augment the sound design of a sequence. The sounds of daily life aren't always as dramatic as a director would like them to be, so the idea of sound design is to construct an audio environment that matches the emotional content of a film. This has been done since the earliest days of modern film production, and is such an effective technique it's still widely used today, even by people who shoot DV (for more detail, see Chapter 17). At the same time, recording images and sound separately adds layers of technical complexity to a production.

If picture and audio aren't recorded at precisely the same speed, they won't look right when played back for an audience. Slight differences in recording speed become noticeable right away, and the difference becomes that much more pronounced over the length of a shot—the longer a shot goes on, the further the audio and picture will drift apart. If a shot contains people speaking and the audio and images are recorded at different speeds, the dialog won't match the lip movements, and even the most inattentive audience will notice. (It is also much more noticeable in close-up shots than from farther away.) To prevent recording and playback problems, film cameras and the audio recording devices they're used with are synchronized (also called synced, also spelled *synched*, both are pronounced "sinked"), so their recording speed remains constant and the material they capture remains in sync.

Synchronized sound, also known as sync-sound, is an integral element of film production and has been since silent films were replaced by "talkies" years ago. In many ways, the impediment to sound in motion pictures wasn't really the difficulty of recording audio but the difficulty of keeping the audio in sync. (Thomas Edison invented the phonograph in 1877, and it became widely available in the early part of the 20th century, but wasn't suitable for use in movies, because it didn't allow for synchronization with film.) Even after the introduction of sync-sound in motion pictures, it would be decades before portable sync-sound production equipment became available. Location sound recording was indeed possible, but it required large and heavy equipment that made the process much more laborious and much more expensive.

Sync-Sound Field Recording and the Birth of Cinema Verite

In the early 1960s, engineers developed portable sync-sound technology that forever changed the way people make films, and set the stage for the small crews and mobile filmmaking style employed by today's DV filmmakers. In 1962, a Swiss company called Nagra introduced a portable reel-to-reel tape recorder with pilot system that allowed it to maintain sync with a film camera. The pilot used crystal sync technology—the Nagra recorder and the camera each contained a quartz crystal that resonated at a precise and identical frequency, and allowed both devices to operate at the same, constant speed (the idea is similar to the way a quartz crystal helps a wristwatch keep accurate time). Because the crystals don't need to be physically connected, the camera and the Nagra operated effectively without being connected by wires, allowing the cinematographer and sound recordist to work in the field without being tethered to each other by cables.

> **NOTE**
>
> *Crystal-synced shoots still required filmmakers to slate each take, and then manually align the picture and audio in post-production (see "Maintaining sync" in Chapter 8). Because it provided an affordable way to ensure image and sound would be recorded at the exact same speed, the technology made sync-sound recording much more accessible.*

Once mounted in small, inexpensive 16 mm cameras like the Bolex (Figure 7-1), crystal-controlled pilots enabled a filmmaker to be truly mobile. (Independent producers still love the Bolex because it's affordable and easy to use. Iara Lee used one to shoot observational material for *Modulations*.) With the advent of affordable, portable crystal sync recording, two or three people could go anywhere and make a professional quality sync-sound production without the expense of a large crew or cumbersome equipment.

Figure 7-1. A 16 mm Bolex camera and an older model Nagra reel-to-reel audio tape recorder.

Technological developments had simultaneously reduced the costs of making a film while providing filmmakers with greater freedom than ever before. The advent of 16 mm sync-sound production helped give rise to new styles of filmmaking, both in documentary (see the sidebar, "Verite and Television, Then and Now") and in fiction film as well. The French New Wave, or la Nouvelle Vague, developed at the same time as portable sync recording and was, at least in part, enabled by the technology. Led by directors François Truffaut, Jean-Luc Godard (I mention Godard's recent film *Notre Musique* in Chapter 17; he also directed *À bout de souffle*, or *Breathless*, in 1960 and *Bande à part*, or *Band of Outsiders*, in 1964), Claude Chabrol, and Alain Resnais, among others, the New Wave made extensive use of location shooting using available light (as opposed to lighting added specifically for the film) and audio recorded in the field (as opposed to studio recordings made under more controlled conditions).

Verité and Television, Then and Now

Portable sync technology gave rise to a new type of filmmaking called direct cinema, or cinéma vérité, which means "cinema of truth" in French. The idea of verite filmmaking is that because the result is a sync-sound film capturing events as they actually happen in front of the camera, it represents a direct slice of life uninfluenced by the point of view of its creator. (The amount of actual "truth" in inéma vérité has since been debated—any project requires interpretation and editorial decision making, both of which undeniably shape its content.)

Some of the earliest filmmakers to experiment with cinéma vérité worked in the newly developing medium of television, and their projects quickly began making their way onto the airwaves. Television networks had always used sync-sound equipment to create their programming, but the equipment was limited to the studio because it wasn't portable. Cinéma vérité, aided by portable audio recorders such as the Nagra, enabled television producers to create shows in the real world and bring them to the screen. While there are significant differences between the following works and shows like *Survivor,* none would be possible without portable sync-sound technology:

- *The Chair,* Robert Drew (1963), chronicles the last days of a death row inmate on his way to the electric chair. This work created the "will he/won't he" genre of documentary. The inmate and his attorneys attempt to stop the execution before it's too late, while the audience watching the film waits to see if they'll be successful or not.
- *Don't Look Back,* D.A. Pennebaker (1967), documents Bob Dylan's 1965 tour; and *Lonely Boy,* Wolf Koenig

and Roman Kroitor (1962), documents the life and work of a young Paul Anka. These films created a new genre chronicling the world of pop stars on and off stage. Long before *Behind the Music,* these two films brought audiences backstage while thousands of teenage girls screamed outside. *Don't Look Back* also contains a musical sequence (a silent Dylan holds up hand-written signs containing his lyrics instead of singing them) that has since been appropriated in music videos by acts ranging from the '80s rock group, INXS, to the hip-hop artist, Common.

- *Law and Order,* Frederick Wiseman (1969), which among other things showed a police detective choking a prostitute, shocked audiences with its portrayal of law enforcement from the point of view of the officers it follows. This film clearly brought audiences into environments most never would have seen, and in many ways, paved the way for current popular TV shows such as *Cops.*
- *An American Family,* Alan Raymond, Susan Raymond (1973), was a television documentary series following a family named the Louds through their domestic travails and an eventual divorce. One child came out as gay to his shocked family, and the wife asked for a divorce—all as America watched. This show was so novel that, even though the producers identified the Louds as actual people, reviewers still wrote about them as characters and suggested ways the show's writers could make them more interesting. This show is easily the progenitor of modern reality television from *The Real World* on.

For the first time, filmmakers could transport audiences into realms too intimate for a large crew, such as a person's home, or locations too remote to accommodate a large production, such as the depths of a rainforest or the top of a mountain. The results were very similar to what we see in DV filmmaking today—people who could never before afford to make films became filmmakers, and people who were already making films tackled subjects that were previously outside their reach.

The Impact of the Camcorder

Because video records sound and image onto the same tape, maintaining sync becomes much less of an issue than in film production. If you're recording audio with a camcorder's *onboard*, or built-in, microphone, you're pretty much guaranteed audio that syncs up without issue, though the audio quality will likely suffer, as the mic can pick up the camera motor sound and might be too far from the subjects to sound great. If you use external microphones to get better sound recordings, you can also record their audio directly to your video cassette using a variety of methods described in the next chapter. Unless you run into real problems, you won't need to worry about your audio falling out of sync, because the same device is capturing both picture and sound. (Chapter 8 explores techniques for maintaining sync when recording audio to other devices.)

In addition to the benefits of sync, because DV tapes are cheap and easy to change, you can leave the camera on and continue to record sound even as your cinematographer experiments with different framings or walks from one part of the room to another. If you're shooting a documentary, this can be invaluable, because the best moments often happen when you aren't expecting them, and if you record good audio, you can piece together a strong sequence even if you miss some of the visual action.

Another benefit of digital camcorders is that they record digital audio at a higher level of sound quality than a CD, which means that unlike material you capture with analog audio devices, the digital audio recordings you make with your DV camcorder won't degrade in postproduction. Because your digital audio and video is stored on a tape as digital information, once you capture the material to your hard drive, you can edit, copy, and transfer your audio from one computer to another the same way you would transfer any other file. This gives you a tremendous amount of latitude, especially when you compare the process with analog video editing in the not-too-distant past—if you were editing on analog equipment, the audio would start to degrade after the first generation, or the first copy you made from the original. In digital postproduction, you can capture audio and video to your editing system without generational loss. Once you've captured your material, you can then create duplicate copies of a file that are identical in technical quality to the original. (Depending on your hardware setup, you may notice some visual or sound degradation if you repeatedly transfer the same footage between an editing system and a camera or deck. Generally speaking, you won't need to transfer material that many times when you're working on a nonlinear digital project.)Also, because good digital recording shouldn't introduce system noise, the recordings are generally clearer. Analog audio, in contrast, is a physical process. The mechanics of audio recording always introduces system noise, or background hiss, to analog audio recordings, which becomes particularly noticeable when material is played back at louder volumes. If you need to, you can also make digital audio recordings louder without losing quality, something you could never do with analog tape because of the system noise (if you make an analog recording louder, the system noise gets louder, too).

Current Recording Options

Depending on what you're shooting, and depending on your working style as a filmmaker, you have a variety of recording options available. The remainder of this chapter explores a number of different routes you can take, and reasons why you might want to choose one over another. Chapter 8 provides further technical details, and describes how you can put the equipment to use and get the best recording possible.

Using a camera's onboard mic

Just about all digital video cameras ship with an onboard microphone, usually mounted at the front of the camera above the lens (Figure 7-2). These mics often record stereo audio, and they don't require any special setup—as soon as you start recording video, the camera mic records audio onto your tape unless you specify otherwise.

Pros

Recording sound with the onboard mic is probably the easiest way to capture audio. If you're in a hectic situation, or working by yourself, this is a no fuss, no muss method that generally provides good results.

Using the onboard mic draws less attention to you during filmmaking than some of the more elaborate audio setups described later. You may want to blend into a crowd (for example, if you're shooting on location and don't have a permit, or if you're making a fly-on-the-wall type documentary), and if you're working by yourself with just a small camcorder, you look like an ordinary person.

A recording you make with the onboard mic is recorded directly to your DV tape, making postproduction a much simpler process. Some of the other methods in this section can get more complicated, especially when it comes to maintaining sync.

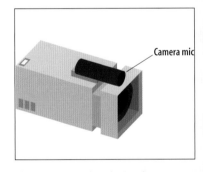

Figure 7-2. An onboard microphone.

Cons

Because the onboard mic is mounted at the front of the camera in a fixed position, the sound recording is always focused in the same direction as the camera. There are often times in a film where important audio comes from another location, for example, behind the camera or off to the side. In these cases, the sound recording won't be as clear as if you had used one of the methods described later.

Sometimes the sound you want to record will be far from the camera. For example, if the focus of your shot is a couple speaking softly on the other side of the room, you might not get a usable recording because the source of the audio may just be too far from the mic. In this case, you might be better off using an additional microphone.

Also, as mentioned earlier, if you're using an onboard mic, your recording may also inadvertently capture the sound of your camera's motor, which would distract the audience from the audio you're trying to record.

Connecting a microphone directly to the camera

DV cameras enable you to connect an external microphone and record sound directly to your tape (Figure 7-3), bypassing the onboard mic. Different cameras use a variety of inputs, and some cameras require an adapter to connect professional audio equipment. Newer prosumer cameras come equipped with professional audio inputs (explained in the sidebar "XLR versus mini" in Chapter 8); check your camera's manual for more details.

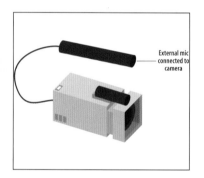

External mic connected to camera

Figure 7-3. A camera with an external mic attached.

Pros

Using an external microphone enables you to choose what type of mic you use. Particular microphone types have different characteristics that are better suited to some situations than others (see "Microphone types" in Chapter 8).

Separating the microphone from the camera enables you to position the microphone to capture the best sound recording possible. Depending on the shot, this may be near the camera or far away.

Because an additional mic is not mounted to the camera in a fixed position, you can angle the mic to record the sound you want, and minimize other sounds that you don't want on your tape. (See the sidebar "The pickup pattern of a mic" in Chapter 8.)

Cons

This is more than likely something you can't do by yourself. Using an additional mic to record good sound requires a person's full attention and is really hard to do when you're also running the camera.

You need to buy (or rent) additional equipment. Depending on the type of equipment you buy, you might pay less than $100 for a basic no frills microphone, or several thousand dollars to buy a good quality mic and the other equipment that goes along with it. If you're already investing your time and money to make a professional quality film for theatrical release or a television broadcast, it makes sense to purchase additional audio equipment (sometimes there's no way around spending money on your work), but at the same time, I don't know too many independent filmmakers with extra cash burning a hole in their pockets.

SIDEBAR

Making Your Camera Battery Last

Whether you're using the camera's onboard mic or a separate mic, a camera will sometimes record a 60-cycle hum when plugged into a wall. Running your camera from the batteries rather than plugging in is a good way to avoid this. It also means that you need to keep your batteries charged.

Older, non-lithium camera batteries last longer if you let them completely discharge (meaning run completely run out) before charging them again. If you recharge a battery before it's completely empty it can develop what's called a memory, and ultimately lose its ability to hold a full charge—you may have noticed the same can happen with the battery on a mobile phone. To prevent this, instead of charging a partially depleted battery, filmmakers often leave a camera on and let the battery run down until it's completely exhausted before charging it. Newer camcorders ship with lithium batteries that are designed to avoid this type of memory effect, so you don't need to let them run down before you charge them. (Even when I'm using newer equipment, I still generally try to completely discharge my batteries before charging them again—call me old fashioned.)

Also, if you're not planning on shooting for a few months, it's a good idea to let the batteries wear all the way down, and then recharge them before you store your camera. Batteries last longer if you don't let them go more than about 60 days without completely cycling.

Recording audio to your camera using additional microphones and a sound mixer

For even greater control over the sound in your film, you can route the audio through a sound mixer (Figure 7-4). A sound mixer is a device that enables you to monitor the audio signals from your microphones and make adjustments before the sound gets recorded by the camera. Using a sound mixer enables you to record with multiple microphones and enables you to adjust each one separately.

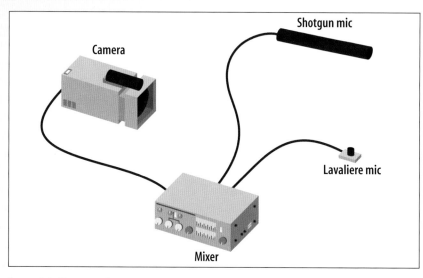

Figure 7-4. Two microphones attached to a camera via a sound mixer.

Pros

A sound mixer, such as the Shure FP-33, enables you to connect up to three microphones to a camera and separately adjust the sound levels of each one (see the sidebar "The importance of audio levels"). While cameras allow you to adjust the sound level of a microphone's input without using an external mixer, using one enables you to make changes without touching the camera—something your director of photography will no doubt appreciate.

Digital cameras record *stereo* audio, which is essentially two recordings —one plays back in the left channel and the other plays back in the right channel (mono audio, in contrast, plays a single recording in both channels). Using a Shure FP-33, you can connect three microphones to a camera and control which microphone's signal is recorded to each channel. This technique is explained in more detail in the section "Using two or more mics, with a mixer" in Chapter 8. A mixer is especially helpful when using different types of microphones together.

Cons

Working with an additional piece of equipment complicates your shoot. Using a sound mixer takes practice. Listening to audio recorded by multiple microphones and making adjustments on the fly takes great concentration and probably won't be the easiest thing you ever do (although the results are worth it if you persevere through the learning curve).

The audio you hear if you listen to the sound mixer may not be the sound the camera is recording. Sound mixers have separate outputs for headphone audio and for the audio that goes to the camera. If there's a problem with the connection between the mixer and the camera, or if one of the settings on the camera has been improperly set, everything may sound great in your headphones but lousy in the final recording. A better solution is to connect your headphones to the camera's headphone output—this enables you to listen to the sound that's being recorded to tape. This may be an inconvenience since you have to stand closer to your DP, but it ensures you know what your tape sounds like. (Higher-end cameras have an audio monitor output that connects to the mixer so you can monitor the actual camera sound, but these cameras are well outside the prosumer price range). You can use headphones with a long cable, but make sure no one trips on it and knocks the camera over, and also make sure the cable doesn't jerk the camera.

Good field mixers aren't cheap: the FP-33 retails for more than $1,250. They're also very heavy, and get even heavier as the day wears on.

Recording to a device other than the camera

A camera is clearly not the only device capable of recording digital audio for your film. Filmmakers often use separate recorders such as a digital audio tape (DAT) deck (Figure 7-5), a mini-disc recorder, or a portable hard disc recorder to capture sound, and then later, in postproduction, combine the sound and picture (very similar to the way people have historically worked in film).

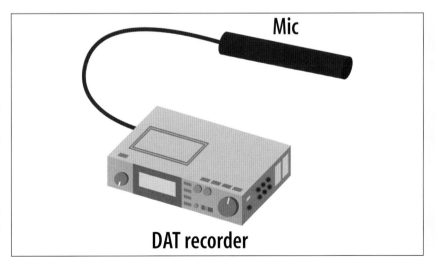

Figure 7-5. A mic attached to DAT recorder.

Pros

The sound recordist and camera operator don't need to be tethered to each other by cables, because the audio is recorded by two separate machines.

What you hear is what you get. A sound recordist listening to the DAT deck or hard disc recorder can listen to the actual recording, and know with confidence what she has on tape.

Many DAT decks and hard disc recorders offer level controls and other adjustments for multiple microphone signals, eliminating the need for an additional sound mixer.

Some recorders can record more than two channels, which might be useful in a shoot with complicated audio, like an orchestra, or five people speaking in a panel discussion.

Cons

These units are not cheap. Depending on the features you're looking for, DAT decks and portable hard disc recorders range from $1,250 at the low end of the prosumer scale, up to more than $4,250 and well beyond.

Like the portable sound mixers discussed earlier, they also require specialized knowledge, both to operate in the field, and when you use their recorded audio in postproduction.

When you record to a separate device, sync becomes an issue. There are ways to ensure that sync doesn't become a problem (some are described in the next chapter), but nothing is as simple as using the camera's onboard mic, or plugging a mic directly into the camera.

In the end, recording good audio is just as important to your film as recording good images. Anyone who knows me has heard me say this over and over again (yes, I sound like a broken record), but it can make the difference between a strong film and something people forget as soon as the lights come up. The next chapter explores the details of prosumer audio equipment, including different microphone types and how you can use them to get the best possible recording in a variety of circumstances.

Digital Audio Production Techniques and Strategies

8

Shooting a good film is all about leaving yourself options, and recording the sound is no exception. No matter how well you plan out a shot, you may come up with a better idea once you get back to the editing room. If you shoot with an eye (and ear) toward flexibility, and record your audio accordingly, you can simultaneously avoid technical mistakes and leave yourself room to take advantage of new ideas as they come to you later on.

A few years ago, I walked by a man sitting outside his apartment building on a summer night wearing a baseball cap that read "Woulda, coulda, shoulda." It's really easy to feel that way when you sit down to edit. If you leave yourself options, especially when you record your sound, cutting your film together can be a much happier experience.

Microphone Types

Different types of microphones record sound differently. Using different mics in combination is an excellent way to leave yourself options—if you record the signal from each mic separately, you can then choose which recording sounds best, and best fits the tone of your story. The first step is understanding the different kinds of microphones available to you as a filmmaker.

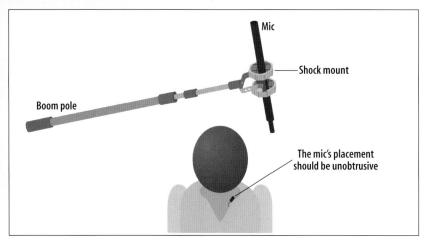

Mic

Shock mount

Boom pole

The mic's placement
should be unobtrusive

Figure 8-1. A shotgun microphone on a
boom, and a lavalier attached to the inside
of a person's shirt collar.

In the world of film production, microphones fall into two broad categories: *lavalier* and *shotgun*, shown in Figure 8-1. Lavalier mics are small, unobtrusive devices that attach to a person's body to capture dialog. The art of using lavalier, or *lav*, microphones entails placing one close enough to a person's mouth to clearly capture what he's saying, while hiding the mic so it doesn't become a noticeable part of the shot, which would distract the audience.

Audio should be so good that no one notices. People always comment on good camera work or visual effects, but notice the audio only when there's a problem, or when a mic sneaks into the frame.

Shotgun mics are designed to pick up sounds from far away and are often placed on a *boom*, or pole, and held over the action just outside the frame. Older models actually did look like shotguns, which is how they got their name.

SIDEBAR

The Pickup Pattern of a Mic

Different microphones have different pickup patterns, or areas in which they optimally record audio. There are three types of pickup patterns—cardioid, hyper-cardioid (sometimes called super-cardioid), and omni-directional. As the name would suggest, a cardioid pickup pattern is heart shaped. If you picture a heart shape extending in front of a microphone, that would be the area in which a cardioid mic does its best recording. A hyper-cardioid pickup pattern is basically a heart shape stretched along its vertical axis, so the area of optimal recording becomes longer and narrower.

All cardioid mics still record some sound from the sides and from behind, but the best and most clearly audible sound is recorded within the heart-shaped area in front. Cardioid mics produce good recordings in noisy environments, because their pickup patterns record sound from isolated areas. Hyper-cardioid mics are especially good for this; you can record clearly articulated dialog from far away while lessening the amount of unwanted environmental sounds or other audio that might distract your audience.

Shotgun mics generally use a cardioid or hyper-cardioid pattern. Lavalier mics often use omni-directional patterns.

Omni-directional mics have a circular pickup pattern and record audio from all directions (the clearest recordings still come from the area directly in front of the mic). The handheld mics you see people use on TV news programs are often omni-directional, because they're easy to use in hectic situations, such as a live news event. Handheld mics like these are not often used in film or video production, because they have to be very close to the audio source to produce a good recording, and would be noticeable in a shot.

The area opposite your mic's pickup pattern is sometimes described as a null area, because it does not effectively record sound. You can use this to your advantage and position your mic to record the audio you want, while minimizing the presence of other sounds in the environment (see the section "Exploit the null" later in this chapter).

Using a boom and a lav together

Because lavalier microphones function differently than shotguns, the recordings they produce don't sound the same. If you simultaneously use a lav and a shotgun to mic an actor reading lines of dialog, the recording from each microphone will have very different qualities. The lavalier, because it's closer to the actor's voice and records sound from a smaller area, will deliver a very clear recording of that actor's dialog. If there were another actor standing nearby, the second actor's dialog would not be as clear in the recording because she was farther away. The shotgun would also deliver a clear recording of the first actor, however, because it records sound from farther away it would also record other sounds—such as traffic on the street outside, or the second actor.

Reality TV shows often mic people with both a lavalier mic and a boom-mounted shotgun. (I'm not advocating the cinematic or moral value of these shows, but their technical production is very good, especially considering the circumstances under which they're sometimes shot.) A person using a sound mixer monitors the volume levels from each mike, and makes adjustments as needed. You can easily work the same way with your prosumer camera. The next section contains a number of sound recording techniques you can use to keep your options open—this way, you'll never have to say "Woulda, coulda, shoulda."

Using two mics, without a mixer

Digital video cameras record stereo audio, which is also called two-channel sound (one channel for the left, one channel for the right). Because they record two channels, prosumer cameras have separate audio inputs for the left and right channels (Figure 8-2 shows the two XLR audio inputs for the Sony PD-170). If you're working with two microphones on a shoot—for example, a lav and a shotgun—you can simply connect one mic to each input. The camera will then record the sound from both mics, separately,

Wireless or Cable Connections

Lavalier microphones connect to other audio equipment either through cables or through a wireless radio system that, essentially, works like a cordless telephone. Logistically, wireless mics are easier to work with because there are fewer cables to trip over. In terms of sound quality, however, wireless mics are highly susceptible to interference from other radio signals in the area, and even from electromagnetic fields generated by electrical equipment. When I was in high school, I had a girlfriend who had to get rid of her cordless phone because it was picking up radio transmissions from a taxi dispatcher (they would wake her up in the middle of the night, even when the phone was turned off). Likewise, just about every audio recordist I know has some type of story about interference on a wireless mic. (There's a good joke about this in the movie *This Is Spinal Tap*, involving a wireless guitar rig.) This doesn't mean you shouldn't use one, just be careful (if you can, try a few tests in the location where you'll be shooting), monitor with headphones as you go, and bring a non-wireless mic with you just in case.

Figure 8-2. Two XLR audio inputs for the Sony PD-170.

to your DV tape. In postproduction you can use the recording that sounds best, or use pieces of each and edit them together.

Using two or more mics, with a mixer

As described in Chapter 7, a portable sound mixer enables you to monitor sound during recording and make adjustments as needed. In addition to controlling volume levels, mixers let you control the *pan* of an audio signal. A pan control enables you to send an audio signal to the left or right channel of a recording, or to both. As a result, if you're working with two mics and a sound mixer, you can send one mic to the left channel and another to the right by adjusting their pan controls.

Figure 8-3 shows a Shure FP-33 sound mixer in which the pan for input 1 has been set to the left channel and the pan for input 2 has been set to the right channel. The FP-33 has two audio outputs, which can each be connected directly to an audio input on a prosumer camera. If the mixer illustrated in Figure 8-3 were connected to a camera, a microphone connected to input 1 would be recorded to the camera's left channel, and a microphone connected to input 2 would be recorded separately to the camera's right channel.

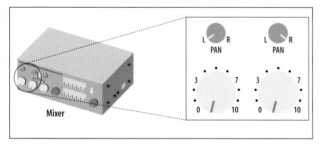

Figure 8-3. A mixer set to record audio to different outputs.

You can also combine signals from different microphones and mix them into a single recording. This enables you to work with more than two mics at a time. For example, if you have three microphones connected to a mixer, you can mix the three signals down to two channels that you then send to the camera. In Figure 8-3, the pan control for input 3 is set to the center, so a mic connected to input 3 would be recorded to both channels. Keep in mind, when doing this you have to live with the results of your mix—you can't undo a sound recording if you don't like the way it comes out. For more control, you'd have to use a multi-track recorder (see "Recording to a device other than the camera" in Chapter 7) and then sync the sound in postproduction.

Maintaining Sync

As discussed in Chapter 7, audiences notice immediately when sound falls out of sync. As a result, maintaining sync is a priority during all phases of production and postproduction.

If you record audio to your DV tape, whether you record directly to the camera or route the audio through a sound mixer, the camera keeps the audio in sync with the picture by using *timecode*. Timecode identifies the exact location of any part of a recording in terms of hours, minutes, seconds, and frames. There are approximately 30 frames per second in NTSC video, so NTSC timecode lets you pinpoint a part of your recording down to 1/30th of a second. DV cameras automatically add matching timecode to the audio

XLR Versus Mini

Professional microphones connect to recording equipment through balanced inputs, also known as XLR inputs. Consumer audio products use unbalanced inputs, also known as mini inputs. Balanced inputs deliver a better quality signal but are more expensive. Mini inputs are more fragile and can easily be damaged during a shoot. Mini plugs can also pop out during a shoot, because they don't lock in place the way an XLR connection does. (The headphone connection on a Walkman or iPod is a mini connection; if your headphones pop out when you're relaxing and listening to music, it's no big deal, but if a connection comes loose during a shoot, it can ruin your recording.) Plugs and inputs (also called jacks) are often described as male and female, respectively. A male plug fits into a female input. (I once explained this to a female colleague who responded that sexism exists even in the naming of equipment.)

Newer, higher-end prosumer cameras (such as the Sony PD-170) ship with balanced inputs. If you have an older camera, such as the Cannon XL-1, or a slightly less expensive model, such as the Cannon GL-2, you need to purchase an adaptor to connect your camera to professional audio equipment.

XLR plug and mini plug

XLR adaptor on Cannon XL-1

and video that you record, and when you capture your footage to a digital editing system, you capture the timecode along with it.

In a previous era of film production, before the introduction of timecode, editors had to manually line up separately recorded sound and image and look for visual cues to check for sync. For example, an assistant editor would check to see if recorded dialog matched an actor's lip movements, and if it didn't, she'd realign the picture and sound until they matched up. Today, digital editing systems capture audio and video in sync from a DV camera, and if by some chance your audio falls out of sync during editing, you can just align the timecode of the audio and video clips, and you're back in business (Figure 8-4).

When you record audio to an external device, such as a DAT or portable hard disc recorder, sync becomes an issue. Some recorders capture audio along with timecode generated by the camera, but these recorders cost a lot of money and require cameras with a compatible timecode output (these are generally well outside the

Figure 8-4. This screen capture, from an exterior shot in our chroma key sequence, contains an overlay that displays the timecode for the video and two audio channels.

price range of an independent—prosumer cameras don't offer a timecode output). Maintaining sync with a non-timecode device is still possible, it just takes a little more work.

Syncing non-timecode audio: slating each take

As a child watching movies, I always loved it when someone stood in front of the camera, banged two pieces of wood together, and yelled "action." I thought this was the very essence of cinema, and someday looked forward to doing it myself because it looked so cool.

> **NOTE**
>
> *Timecode appears as hours:minutes:seconds:frames. For example 01:20:15:28 indicates a location 1 hour, 20 minutes, 15 seconds, and 28 frames after the start of a recording. Unlike the tape counter on a VCR, the timecode recorded onto a clip never changes—no matter what deck or editing system you use to watch your footage, the same shot and the same audio will always be at the exact same location.*

In my first year of film school, I discovered that the two pieces of wood are called a *slate*, and banging them together helps an editor sync audio and picture. When audio and picture are recorded separately, an editor can look for the frame of video where the two wooden pieces of the slate first come together, and can align that frame with the start of the sound generated by their impact. The picture and audio that follow will be in sync for the duration of the shot. Once the shot ends, the editor will need to redo the process for any subsequent shots.

You don't need to buy a slate to use this technique (although I must admit, a slate is fun to use). Many filmmakers on a budget have a production assistant stand in front of the camera and clap his hands together once (Figure 8-5). The single clap performs the same function as banging together two pieces of wood, it provides a reference point to sync picture and sound. (Filmmaker Michael W. Dean uses a clap slate in the opening shot of *D.I.Y. or Die: How to Survive as an Independent Artist*). Other filmmakers slate a shot by holding a microphone in front of the camera, and tapping the top of the mic once with their hand—it serves the same purpose. (In *Bright Leaves*, filmmaker Ross McElwee—who also directed *Sherman's March*, described in Chapter 1—recruits his son as a sound recordist and shows him on camera slating a few shots using this last technique.)

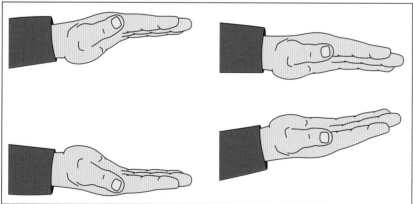

Figure 8-5. This is a basic clap slate. If you were manually syncing an audio recording to a slate like this, you would position the sound of the clap to start where the hands come together. Once you've positioned the clap, the entire shot will be in sync.

Syncing non-timecode audio without a slate

Unfortunately, no matter how well you plan, there will always be shots where slating each take is either impractical or just something you forget to do. (Once, during a shoot, my director of photography and I came up with an elaborate plan to clap slate each shot and unintentionally abandoned it after about 20 minutes—we just got caught up in all the other details of the shoot.)

As described earlier, when you record audio directly to your DV tape, the camera captures audio and video together and keeps them in sync. However, if you're recording audio to a separate device, such as a DAT recorder or a portable hard disc recorder, you may find yourself with audio but no reference slate to sync it up with the accompanying video.

Fortunately, one of the basic elements of digital audio, called the waveform display, can also help you sync non-slated recordings. In digital postproduction, audio recordings are displayed graphically on a computer screen using a waveform. The loudest parts of a recording are represented as peaks in the waveform, and the quiet parts are represented as flat (this means the highest points are also the loudest). With practice, editors learn to read waveforms and can tell where words of dialog begin and end, where background audio peaks and fades, and where to add music. A waveform can also provide a reference for syncing additional audio tracks.

If you've recorded audio to an external device while you were also recording video to your camera, you more than likely captured audio to your DV tape using the camera's onboard mic. When you capture the audio from your external recorder to your editing system, along with the audio and video from your DV tape, you can use the waveform displays to align both of the audio tracks. Even if you recorded the audio separately, using different microphones placed in different locations, as long as you recorded the same audio (a person speaking, a musician playing an instrument, etc.), the waveforms will be similar enough to align your audio tracks. In Figure 8-6, the camera audio's waveform on track A1 doesn't look exactly the same as the waveform for the separately recorded track A4, but they look similar enough to line up the tracks. You can also align separate audio recordings by ear—place each on its own track in the timeline and slide the externally recorded track back and forth until it no longer "echoes" against the audio recorded directly to the camera. Once you've synced up your audio tracks, you can either mute the one that doesn't sound as good or simply delete it from the timeline.

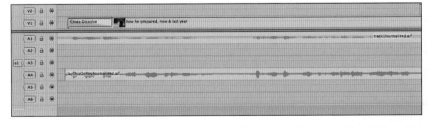

Figure 8-6. A timeline in Final Cut Pro with wild sound aligned to the waveform display of the camera.

If worst comes to worst, and you didn't record either a slate or a reference audio track to your DV tape, you can always sync the audio to your picture manually. Capture your audio to your digital editing system, and capture your video as well. Adjust the start of your audio clip in relation to your video, and play them back together. Carefully watch people's lip movements to see if they match the dialog. If they match, you're all set. If they don't, readjust their alignment and try again. This method takes a fair amount of patience, but it's entirely doable. Once you nail down the sync, your audience will never know the recording wasn't perfect from the start—they'll just watch your movie and get wrapped up in the story.

The Smart Slate

If you really want to make every slate count, and you have the budget, you can use a smart slate. A smart slate has a large digital timecode display on its face, and a built-in timecode generator that jams, or syncs up, to the timecode of an external audio recorder. Using a smart slate helps an editor sync audio and picture in postproduction by not only providing a visual cue (like the methods described earlier) but also by clearly displaying the timecode of an audio recording at the head, or beginning, of the shot.

Of course, to use a smart slate you need an audio recorder that handles timecode. As mentioned earlier in this chapter, timecode audio recorders are generally well outside the price range of an independent producer (a timecode DAT or hard disc recorder sells for more than $4,000). A smart slate itself sells for about $1,500.

Strategies for Recording Good Digital Audio

I live down the street from Katz's deli, where they filmed my favorite part of *When Harry Met Sally*, the "I'll have what she's having" sequence. Director Rob Reiner (who also directed *Spinal Tap*) gave his mother the eternal gift of saying the line on camera, and with those five words, magnificently ended one of the most memorable comic sequences of any movie.

Think about everything that would have been lost if the line hadn't been recorded clearly. The scene would certainly have had comic value without its closing line, but its impact would be significantly weakened if people didn't understand what she said. Hollywood films like *When Harry Met Sally* record take after take of important lines, and often have actors re-voice their dialog in a specially equipped studio if the director doesn't feel the location sound was recorded well enough. Dialog propels a film forward, and if an audience can't immediately understand what actors are saying, big parts of the story get lost. Unlike a person reading a book, who can go back to the top of the page and reread something he didn't quite follow, a viewer watching a movie either immediately understands the words in a piece of dialog, or the lines get lost forever. You can develop a story so a line's true meaning comes to light only as the story progresses, but if the words are too muffled or the audio level is too low for people to hear what someone is saying, the scene you worked so hard to record just won't make the cut.

Audiences know what good audio sounds like. People grow up watching big-budget television shows and going to see Hollywood movies that always sound really good, even when the stories are pedestrian. Television networks and film studios spend lots of money to make sure everything in a film sounds perfect (see Chapter 17), and audiences expect your film to sound just as good—in fact, audiences are often more critical of audio problems than of things like shaky camera work or soft focus. Ensuring the audio in your production holds up to the standards set by the big shots is no easy task, but it is within your ability. This section offers a few techniques to get you headed in the right direction.

Proximity

This first suggestion is very simple—bring your microphone as close as you can to the sound you're recording. As described earlier, if you record audio from a distance, you also record additional sound in the environment. Placing your mic closer to the source helps you get a cleaner recording. You can also use the pickup pattern of your mic to focus on the sound you want. If you're using a cardioid or hyper-cardioid mic, you can concentrate on surprisingly focused areas of audio by placing your mic carefully, and attentively aiming your pickup pattern.

The ideal proximity for a boom mic is just outside the frame, so it's close to the subject but not visible to the camera. Before the cinematographer starts filming, it's often helpful to define the boundaries of the frame by slowly bringing the microphone as close as possible to the subject and having the cinematographer yell out "boom" when the mic becomes visible in the frame. Once the boundaries are defined and recordings starts, the boom operator and cinematographer can work together to ensure the mic is as close as possible but stays outside the frame.

Exploit the null

Use the null area of your mic's pickup pattern to exclude sounds you don't want (see the sidebar "The Pickup Pattern of a Mic" earlier in this chapter). In an ideal world, a filmmaker would be able to work in complete silence and record only the sounds related to her film. This is the idea behind a Hollywood studio spending millions to recreate a house or even a city street on their property rather than just using an actual house or a street that already exists. If, however, you can't afford a soundstage, or if you're shooting on location, there will always be environmental sound to contend with. One way to deal with unwanted sound is to focus your mic's pickup pattern in the opposite direction. This is equivalent to framing out an unwanted visual element when you compose a shot.

For example, if you want to record an actor's dialog, but you don't want to record the traffic sound coming in through the window:

1. Position yourself between the actor and the window and aim the pickup pattern of your cardioid or hyper-cardioid mic at the actor.

2. Angle the mic so that while the pickup pattern stays focused on your actor, the null is aimed at the sound you don't want.

This way, the mic is trained on the sound you want and pointed away from the sound you don't. If you bring your mic close to the subject and exploit the null, you can record excellent audio in noisy locations.

Turn off electrical appliances

Appliances like refrigerators, air conditioners, and even computers generate substantial amounts of noise. While people tune these sounds out in every-day life, they can make it really hard to record clean audio. Before you start shooting, unplug anything that might create unwanted sound during your shoot. (This includes phones, both mobile and land lines—they always ring at the worst possible time, and it's very difficult if not impossible to filter a ring out of a sound recording. Mobile phones set to vibrate interfere with audio recordings when they receive or send calls, the interference sounds like Morse code.) When you finish shooting, make sure you turn everything back on. If you're shooting on location at someone's house and forget to turn the refrigerator back on, all the food will spoil.

Wear headphones

This is so obvious some people forget. Microphones capture audio differently than the human ear. Even if you're standing right next to a microphone, you have no idea what you're recording unless you wear headphones and listen. I've worked with a countless number of students who recorded unusable audio because they didn't wear headphones, and as they quickly discovered, there's just no way around this one. Just like you wouldn't shoot video without looking through the viewfinder or watching the LCD display, there's no reason to record audio without headphones. You can even keep a spare pair of tiny Walkman-style "earbuds" in your camera bag in case you forget your good headphones, or in case they fail on a shoot.

Record a test and play it back before you get into the meat of a scene

Once you've got everything set and you're ready to record, do a quick test. Turn on the camera, record a minute or so of audio and video to test your full setup, and then playback the results. Listen to the audio output of the camera using a good pair of headphones (again, there's no substitute for headphones). If you hear everything you're intending to record, and if everything looks like it's in sync, you're all set. If not, there's no better time to make changes.

If you don't think something sounds right, start over. You've already gone to the effort to assemble a crew, pay for equipment, and make a really big personal investment in the shoot. Why not redo some audio recording?

When everything sounds the way you want it to, you're in good shape.

This chapter marks the end of the first half of the book, which explores ways to record the high-quality audio and video you need to realize your creative vision. At this point, you've read about technical specifications and production techniques in detail. You've read about frame rate and aspect ratio at the beginning of the book, lighting and color temperature, the strengths of microphones and sound mixers, and ways you can use them to your advantage.

The next part of the book, *Digital Postproduction for the Independent*, shows you how to put everything together. Chapter 9 introduces you to the technical principles that make digital editing systems possible, and how they benefit an independent filmmaker. In the chapters that follow, this book explores editing (video and audio), creating effects, and outputting your project to DVD, video tape, or the Internet.

Congratulations. Now, get ready to make the film you've always dreamed about.

An Overview of Nonlinear Editing

9

Just as prosumer DV cameras revolutionized the way independent filmmakers shoot their work, nonlinear editing systems have forever changed the way people edit. For years, people dreamed about the ability to not only shoot with an affordable camera, but to edit on an affordable system without sacrificing quality. When Avid introduced its digital editing systems in the mid-1990s, filmmakers marveled at how easily they could create multiple versions of a project and how, suddenly, it had become so much easier to insert or delete material from a sequence. Almost overnight, rearranging and juxtaposing shots became about as simple as cutting and pasting text in a word processor.

The Avid, however, was not cheap. In fact, at upwards of $60,000 for a complete system, an Avid easily cost more than many independent filmmakers earned in a year. When Apple introduced Final Cut Pro in 1999, professional-quality nonlinear editing became affordable. Final Cut retails for slightly less than $1,000, and unlike an Avid, which required users to purchase proprietary hardware additions like specific capture cards and hard drives, Final Cut runs on standard, off-the-shelf computers.

By the year 2000, filmmakers could shoot high-quality video with cameras that cost approximately $3,000 and edit their work on computer systems that cost about the same (perhaps a little more with extras like a large-screen monitor and some additional memory). Filmmakers working in DV now had access to professional-quality equipment that enabled them to work more affordably than ever before, and with significantly greater efficiency.

> **NOTE**
>
> *What matters most, of course, is your ability to use technology to realize your artistic vision (a friend of mine from film school once compared editing a project to building a house of cards—if you take out the wrong piece, everything around it collapses). Regardless of the technology employed, arranging your content to tell a good story remains an editor's greatest challenge. Just as knowing all the features of Microsoft Word does not in itself make someone a good writer, knowledge of technology does not automatically make someone a good filmmaker.*

Software Options

Following the success of Final Cut Pro, Avid introduced its own software-based prosumer editing tool, Avid Xpress (*http://www.avid.com/products/xpressFamily/*), which is available for Mac and Windows. (Avid also continues to make high-end, high-priced, hardware-based systems, which are used by major film studios and production companies.) Avid Xpress currently comes in three versions:

- The base model, Avid Xpress, retails for slightly less than $700 and works only with DV footage. (According to the Avid web site, *www.avid.com/products/datasheets/xpressdv.pdf,* Xpress DV requires Windows XP Professional to run on a PC and OS X 10.2.4, 10.2.6, or 10.2.8 to run on a Mac—OS X 10.3 and 10.4 were not supported at the time of this writing.)

- Avid Xpress Pro, which retails for less than $1,700, offers a wider range of effects, an expanded suite of compression options, and most importantly, supports media other than DV (if you wanted to edit media shot in a professional format such as Betacam SP or Digital Betacam, you could do so in Xpress Pro but not Xpress DV).

- Avid Xpress Pro HD enables users to work with High Definition footage. (For more information on HD video, see the sidebar "Standard Definition versus High Definition Video" in this chapter and "HD Aspect Ratio and Frame Rates" in Chapter 2.)

In late 2002, Apple released Final Cut Express *(http://www.apple.com/finalcutexpress/)*, which, like Final Cut Pro (*http://www.apple.com/finalcutpro/*), runs only on Apple computers. Final Cut Express is designed for people who want to do more than they can with consumer products like iMovie, but who don't want to spend the money to buy a full version of Final Cut Pro, or spend the time it takes to learn the full array of Final Cut Pro's professional features. Final Cut Express retails for approximately $300, and newer versions work with DV and HD footage. Like Avid Xpress, Final Cut Express is not compatible with standard-definition professional media in formats such as Betacam SP or Digital Betacam (to edit these, you would need to work with Final Cut Pro).

Final Cut Express became a competitor of Adobe Premiere (*http://www.adobe.com/products/premiere/*), which had for years been the standard for affordable desktop video editing. Premiere, which has been around since the early 1990s as a tool to edit analog video and since 1998 as DV editing software, now comes in two versions:

- Premiere Elements (*http://www.adobe.com/products/premiereel/*), released in early 2005, retails for approximately $100, and enables basic DV editing. Premiere Elements is not compatible with standard definition media.

- Premiere Pro retails for approximately $700 and is compatible with a wider array of production formats (including HD and SD). Premiere Elements and Premiere Pro both run only on the Windows platform.

Some digital editing systems enable a filmmaker to output an edit decision list, or EDL, which he can then take to a high-end production facility and, with the help of the facility's staff, create a broadcast-quality version of his project. (EDL creation is described in more detail in Appendix A.) Generally, only people working on larger-budget productions take this approach, because the process can be very expensive.

Final Cut Pro and Premiere Pro enable users to output an EDL. Final Cut Express and Premiere Elements do not. Avid Xpress DV, Avid Xpress Pro, and Avid Xpress Pro HD all support EDL output. (You can also edit a project at home using Avid Xpress, and then take it to a post-house to finish the project on high-end Avid gear.)

Avid Xpress DV, Final Cut Express, and Premiere Elements are all knowledge scalable. This means if you learn to edit on the lower-end versions and then move up to a more expensive version of the same product, your skills will transfer from one version to another.

Timelines, Frames, and Tracks—How Nonlinear Editing Benefits the Independent Filmmaker

The following sections describe the advantages nonlinear editing provides for the independent filmmaker.

Tape-to-tape linear editing

Video editing was not always a flexible process. While 16 and 35 mm film production enabled editors to cut a shot from the middle of a sequence without disturbing the material around it, video was originally linear (see "Linear vs. Nonlinear Editing" in Chapter 1 as well as the section on non-destructive editing in this chapter). Editing linear video entailed placing one shot after another on a tape, in the exact sequence, and at the exact length that would appear in the finished product. There was no button marked "undo." Because each shot appeared exactly after the last, there was no way to go back and make changes to the earlier parts of the tape without ruining everything else that had been added later. This led to high levels of anxiety in the editing suite, because edits were irreversible, and an increased tolerance of mistakes (it was so difficult to go back and make changes that people sometimes left bad edits in a program, even if they weren't satisfied, because it was so difficult to take them out).

In 1995, at my first job editing video, I worked with an analog editing device connected to two decks. (This was a standard operating procedure at the time—if you watch reruns of *Mad About You*, you can sometimes see one of these old editing systems in the background of a shot. The main character is a documentary filmmaker.) I played the footage on deck A, found the section I wanted, and then set the in point to mark the beginning of my selection. Next, I set the out point to mark the end of the shot. I then had two choices—I could hit a button marked "preview" or another marked "edit." Hitting the preview button displayed a safe preview of what the edit would look like but did not commit the actual edit to tape (this left me room to change my mind). When I hit edit, the controller would play the footage on deck A, also called the playback deck, and then record the footage onto deck B, the record deck. Figure 9-1 shows a similar setup.

Standard Definition Versus High Definition Video

High Definition video, also called HD video, is a collection of digital formats that offer higher resolution images and higher frame rates than traditional forms of analog or digital video. HD also employs a widescreen 16x9 aspect ratio that gives images a more cinematic look. (For more details on resolution, frame rate, and aspect ratio see Chapter 2.)

All video formats other than HD are referred to as standard definition video, or SD video. Standard definition video includes all analog video formats, ranging from consumer hi-8 video up to professional quality Betacam SP, and many widely used consumer and prosumer digital formats such as mini DV and broadcast-quality Digital Betacam.

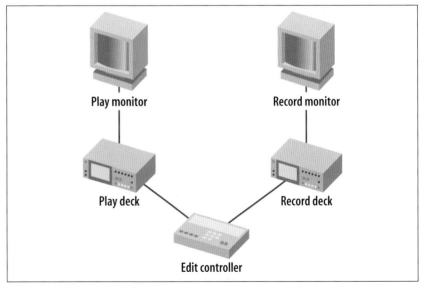

Figure 9-1. A three-quarter inch tape-to-tape analog editing system.

Because I was paid by the hour, my boss specifically instructed me never to hit preview. This was high-stakes editing. Not only because I wasn't previewing my edits, but because analog video editing didn't allow going back to make changes. If I got to the end of a show and then realized something was missing from the beginning, there was no way to insert it without redoing a day's worth of work, or more, to fix it. (I won't even tell you about the things I left out.) Today, to the joy of editors around the world, adding or removing a shot requires nothing more than a few keystrokes or a mouse click.

Nonlinear editing with timelines and frames

Digital editing systems display an editor's arrangement of shots onscreen using a series of frames in a timeline (Figure 9-2). Material at the start of a sequence appears at the left end of the timeline, and later shots appear farther down the timeline to the right. Spaces in the timeline are measured in frames, based on the frame rate of the source material (for example, 30 frames per second for NTSC video or 25 fps for PAL; see Chapter 2).

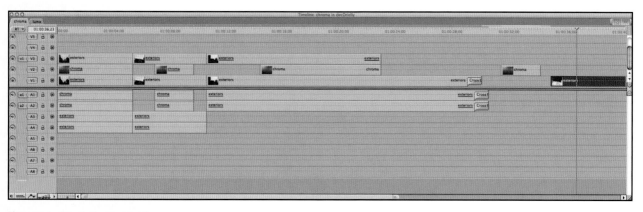

Figure 9-2. Clips in the Final Cut Pro timeline.

If an editor wants a shot to appear earlier or later, she can simply move the shot to another part of the timeline. This provides an editor with infinitely more freedom, and the ability to experiment with different changes without risking disaster. If the editor doesn't like what the change looks like, she can just hit "undo." (This is referred to as *nondestructive editing*, and is explained in the next section of this chapter.)

Timelines also use a series of tracks, arranged vertically in the timeline, which enable an editor to place more than one audio or video clip on the timeline at the same location. Editors often layer audio to create intricately faceted compositions from multiple sounds; for example, the voice of a film's protagonist layered with ambient sound and perhaps some music (see Chapters 17 and 18). A skilled editor can also layer multiple shots of video to create complex title sequences and sophisticated composites (see Chapters 11, 12, 14, and 15).

Making the Most of Nondestructive Editing: The Difference Between Project Files and Media Files

Editing systems like Avid and Final Cut Pro did not provide filmmakers with their first introduction to nonlinear editing. Editing reels of 16 mm and 35 mm film has traditionally been a nonlinear process, but unlike digital editing, it's a destructive one.

Filmmakers often refer to editing as "cutting a project together" or simply, "cutting a project." To a person working on a nonlinear editing system, the idea is metaphorical, but traditional film editing involved physical cutting and splicing. Film editors literally cut a piece of film to select the segment of footage they want (that's where the term *clip* comes from) and splice it into a sequence of other clips. The problem is, after a few cuts, editors quickly find themselves with a room full of very short leftover clips, called *trims*, that are notoriously hard to manage. If you trim say two frames off a shot, you then have to keep track of those two frames, because you might need them later. Editors who physically cut film knew not to throw anything away, because then they'd be stuck if they tried to reedit. Rather than leaving things on the cutting room floor, editors hung strips of film on individual hooks in large bins to keep them organized. At the end of the day, assistant editors spent hours *restoring* trims, or pasting very short clips back onto the longer shots they came from (otherwise, you can only imagine where they might have wound up).

> **NOTE**
>
> *When something is shot, but doesn't end up in the final edit, it is said to "end up the cutting room floor," even sometimes still in video, when nothing actually ends up on the floor. Example, "My performance was great, but that director hates me. I ended up on the cutting room floor."*

The function of project files

Editing film by hand was a labor intensive physical process that could strain the organizational skills of even the most meticulous editor. Today, the process is very different. When you edit a sequence of shots together in Final Cut Pro or another nonlinear application, you're not editing the actual video clips you captured to the computer. Instead, you're creating a series of instructions, telling the computer to play designated portions of the clips one after another. As a result, you have two sets of files; *media files*, which are the actual video and audio clips that the computer stores in their entirety, and *project files*, the instructions you create as you arrange different clips in the timeline (some systems, such as Avid, refer to these as *pointer* files).

The main benefit of programs that use separate media and project files is you can edit without fear of destroying your source footage. If you make an edit you don't like, it's no big deal, your source footage remains unchanged so you can try out different ideas. You can also create multiple versions of a sequence to see which one works the best. You'll never know how two shots work together until you place them in the timeline side by side, and with nondestructive editing, you can experiment without consequence.

Each time I begin a new *pass*, or a new attempt at editing a sequence, I save a new version. I label each sequence numerically, for example classroom01, classroom02, etc., so I can easily identify the newest version (I also save the older versions so I won't be stuck if I don't like the newer changes).

In addition, I periodically back up my project files onto an external disk drive, or onto removable media such as a zip disk. Because project files take up very little memory, a complex hour-long project might require only a few megabytes of storage space (duplicate media files, on the other hand, quickly fill up a hard drive at the rate of one gigabyte per five minutes of video). If you archive your project files, you can easily use them to rebuild your project in the event of a computer crash. Rebuilding a show from project files is infinitely more pleasant than trying to reproduce weeks worth of work, and infinitely more practical than saving duplicate versions of an edited show with all its media files. (Jan L. Plass, who chairs the doctoral program in Educational Communication and Technology at New York University, jokes that there are two categories of computer users: those who have lost everything on their hard drives, and those who will.)

How Timecode Makes Nonlinear Editing Possible

Timecode, which was briefly introduced in Chapter 8, identifies each frame of audio and video in your project. Timecode measures a tape in hours, minutes, seconds, and frames, placing a unique timestamp on each frame of footage. This means that if you're shooting video at a standard 30 fps, timecode can accurately identify the location of recorded material down to 1/30th of a second. Timecode appears in a standardized format, written as hours:minutes:seconds;frames. For example 01:15:20;05 means 1 hour, 15 minutes, 20 seconds, and 5 frames from the start of the tape. See Figure 9-3.

Figure 9-3. The timecode field in the Final Cut Pro timeline.

There are two timecode formats: drop frame and non–drop frame. NTSC video, the North American broadcast standard, uses drop-frame timecode. Drop-frame timecode is made to conform to a broadcast hour and, using a complicated series of calculations, periodically drops a series of frames to run at 29.97 fps instead of an even 30 fps. Drop-frame timecode is written with a semicolon separating the seconds from the frames (HH:MM:SS;FF); non–drop frame timecode, as in the example in Chapter 8, uses all colons (HH:MM:SS:FF).

In the past, independent video producers often used non–drop frame time-code, because it was very hard to calculate which frames to drop, if they weren't using a computer. Digital editing systems now perform drop-frame calculations as part of their general functions, so working with drop-frame timecode is not only a simplified process, it also makes your show broad-cast friendly in the event you're invited to screen your work on television, because drop-frame is the standard broadcast television frame rate.

As described in Chapter 8, timecode helps keep audio and video synchro-nized during recording and editing: when the timecode of a video frame matches the timecode of an audio frame, they're in sync.

Timecode also enables the *random access* that makes nonlinear editing pos-sible. Because timecode uniquely identifies each frame of footage in your project, you can instantly access any part of your captured material just by typing in its timecode. Instead of slogging through hours of footage to find a 20 second shot, timecode enables you to navigate to the precise location of the shot's first frame. The ability to identify and find a single frame is described as being *frame accurate*. If you're working with a 12:1 shooting ratio, meaning you shoot 12 minutes of footage for every one minute that winds up in your finished project, you can imagine how much tape you'd have for a two hour film and how much time you save by navigating directly to the frame you want (see the sidebar "All about the ratio" in Chapter 1).

Timecode information doesn't change. The timecode on footage you record with your camera stays the same when you capture the footage to your computer's hard drive, and still stays the same when you edit the material into a sequence. In addition to making nonlinear editing possible, this con-sistency enables nondestructive editing. Earlier I described project files as a series of instructions telling an editing application to play particular sec-tions of various media files in a designated order. Timecode is what enables your editing application to identify each section of the media files. No two timecode locations on any tape are alike, and even though two tapes can contain the same timecode, editing applications allow you to name each individual tape to avoid confusion. (Chapter 10 explores techniques to help you mange your footage and prevent problems.)

As a result, the project files for your film are essentially a list of the begin-ning and ending timecode locations of all your shot selections, arranged in the order you specify and updated every time you make changes. Once your project is finished, you can use this timecode information to create an EDL, as described earlier, to create a high-quality broadcast master. An editor who shoots film can also use timecode information to strike a film print for theatrical release by matching the timecode of material in the master to *keycode* information on the original negatives. Both processes are described in more detail in Appendix A.

In the past, editing an independent film or video project either meant pay-ing studio rental fees, or working in the middle of the night while somebody you knew wasn't using his editing suite—digital editing systems were simply too expensive for most individuals to purchase. Today, buying the equipment to build your own postproduction studio is not only a possibility but, in many cases, proves more economical than renting. Once you make the initial purchase, the cost-per-use of your editing system lowers each time you sit down to edit. Depending on the number of projects you work on and the payment terms you negotiate with your clients, your digital editing system can even become a revenue source. (If you work the balance correctly, you can not only do what you love, but you can make a living at it—how great is that?) Chapter 10 walks you through the process of configuring your own studio and gets you ready to edit.

Setting Up Your Digital Post Facility

10

Assembling your own digital editing system makes you the boss. You can edit on your own schedule, without competing with other people who have projects to finish, or running up your credit card bill to rent time at someone else's editing suite. This cost-effective flexibility not only saves you time and money, but is likely both easier and more affordable than you may think.

Putting a digital studio together entails setting up a computer equipped with editing software, connecting your computer to a deck or camera to capture audio and video, and configuring a hard drive to store and organize your media. Lastly, because computers and televisions display images differently, it can also be helpful to connect the output of your editing system to a dedicated NTSC video monitor. Whether you're buying a new computer and building your system from scratch or optimizing a computer you already own for use as an editing station, the speed of your computer (both in terms of its *processor* power and the amount of *random access memory*, or *RAM*, you install) plays a central role in the success of your system.

Is My Computer Fast Enough?

A friend and fellow author/video editor named Bonnie Blake once compared purchasing a new computer and not installing additional RAM to buying a fast Mercedes and not filling up the gas tank. Her point was that even a brand new computer equipped with a very fast processor still needs a substantial amount of RAM to perform at its best. Editing video will tax the resources of an exceptionally powerful computer. (Most computers are not used to edit video, so most computers do not ship with enough RAM to do so.) Working with a dual-processor computer and installing additional RAM will enable you to work more quickly, and to take advantage of real-time effects playback and other new software features in digital editing programs like Final Cut Pro.

SIDEBAR

How Do I Find Out About My Computer's RAM and Processor?

To check the amount of RAM installed on a Mac, and to see what type of processor the computer is using, use the About This Mac function.

1. Click on the Apple icon in the upper-left corner of the screen, and hold the mouse button down. A drop-down window opens.

2. Choose About This Mac from the drop-down menu, and release the mouse button. A new window appears, displaying the amount of memory (RAM) installed on your computer, along with the type of processor and the operating system version.

To check the amount of RAM installed on a PC, and to see what type of processor the computer is using, use the My Computer icon.

1. Right-click the My Computer icon on the desktop. (If the My Computer icon is not displayed on your desktop, go to the Start menu, and right-click the My Computer icon in the Start menu.)

2. Select Properties. A new window opens.

3. The bottom of the General panel displays the processor speed and the amount of installed RAM.

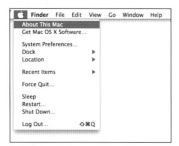

NOTE

The RAM in a computer functions in a way that's very similar to a person's short term memory. Computers temporarily store information in their random access memory and use the information when they perform calculations. For example, if you roll your cursor over an image on a web page and the image changes, the computer uses RAM to store both states of the image and to change from one to the other in response to your actions. Similarly, if you cut a paragraph of text from a document or a video clip from a sequence, your computer uses RAM to store the information in its memory until you paste it somewhere else.

Audio and video playback, and especially editing, require significant amounts of processor power. Computers with faster processors, or a set of dual processors, perform better because they can handle more information. If, in the course of running through the series of functions a computer performs every second, it finds itself overwhelmed with a task, the computer pauses to finish that task before moving on the next. For example, a slower computer might hang up or momentarily freeze while playing back a video clip, because the processor gets overwhelmed.

A computer's processor is not something you would ordinarily upgrade after purchase, but you can easily augment your computer's performance by installing more RAM. Adding RAM increases the number of tasks a computer can perform at a time. One immediate benefit to installing additional RAM is that it enables a computer to run multiple applications together, speeding up your workflow. Editors building a complex title sequence often work with Illustrator or Photoshop and After Effects at the same time, creating graphics in

Illustrator or Photoshop then importing them into After Effects to animate them. Illustrator, Photoshop, and After Effects all use significant amounts of memory, so an editor working on a computer without additional RAM might have to quit one application before launching one of the others, dramatically slowing down the editor's workflow.

Even if you're working with a single video editing application by itself, the available RAM installed on your computer becomes particularly noticeable, and particularly important, when you create complex sequences and start to add any type of compositing effects. When a sequence contains effects, for example several video streams composited into a montage, your computer needs to render the effects before you can play them back. Rendering a sequence means the computer creates a new series of video clips that contain each frame of audio and video, in addition to all the effects that were added to them. As you can imagine, generating video files, and in particular generating video files that contain effects, is no simple task. Rendering requires all the power your computer can muster. In the recent past, even simple effects took a tremendous amount of time to render, often forcing editors to stop working for 20 minutes or more while they waited. This was great for editors who liked to take breaks while they were still on the clock. When working on a deadline, however, waiting for a computer to render could be a deeply painful experience.

> **NOTE**
>
> *A computer's processor operates in cycles, kind of like a clock. Computers constantly perform a number of functions each second, and cycle through all the functions in order before starting over again with the first. The speed of a computer's processor measures how many times it can cycle through the functions in one second. For example, an Apple PowerBook with a 1.33 GHz processor performs significantly faster, meaning it processes more information each second than a comparable computer with a 1 GHz processor chip.*
>
> *The processor speed number on a PC (Pentium, AMD, etc.) is not the equivalent of the processor speed on a Mac. For example, a dual 1.8 GHz G5 is the equivalent in processor speed to a single 3.3 GHz P4. Additionally, an AMD 64 FX with a 2.2 GHz processor is equivalent in speed to a dual 1.8 GHz G5. The architecture between these systems is different. With a Mac processor, data gets from one point to another more quickly because it involves fewer stages (less work) than a typical PC processor, such as a P4 or an AMD.*

SIDEBAR

Rendering Performance and Hardware Requirements

To enable editors to work more quickly, Apple introduced what it calls real-time effects playback to Final Cut Pro. Real-time effects playback, which Apple also calls RT Extreme, means that an editor working on a fast enough system can preview various effects without waiting for her computer to render them. Real-time playback makes significant demands on a computer's processing power and RAM, as does editing HD video.

Among the system requirements Apple lists for Final Cut Pro editing systems are a computer with a 500 mHz or faster G4 or G5 processor and 512 MB of RAM. Using HD features, however, requires a 1GHz or faster processor and a minimum of 1 GHz RAM. (Apple recommends 2 GHz.)

Even for computers that meet these minimum requirements, there are different levels of RT Extreme capability, depending on the speed and power of different computers: systems with more RAM and faster processors can view a greater number of effects in real time at higher levels of quality.

Regardless of the software you use, any application has basic hardware requirements. If your system doesn't have a fast enough processor and adequate RAM, you won't be able to run the program.

Digital editing software applications list their system requirements on the outside of the box and on their web sites. Be sure your computer meets these requirements—and then some—before you install any new software. Many retailers won't let you return software once the box has been opened, even if the software won't run on your system.

If you're buying a new computer system to use for editing, it's often a good investment to purchase the fastest model available. Even if you can't imagine using all the processing ability of a higher priced computer, you may find that you need the power sooner than you think. As software programs add new features, they require additional power and memory to operate. If you're certain you never want to upgrade your system, it's not an issue (many filmmakers produce great work using older equipment), but if you want to leave yourself the option, it's worth thinking about. A good analogy is a lawyer or brick layer who charges a higher hourly fee but saves you money in the long run by finishing more quickly than a slower competitor who bills at a lower rate. Likewise, it may be cheaper to buy a faster machine now than to buy a less expensive model today and an entirely new system in the near future because the processor is too slow to run the software you want. Keep in mind that slower systems crash more, too—crashing in the middle of a 10-hour render means starting over.

Additionally, computer manufacturers constantly come out with new models that feature newer processors, for example a G5 processor instead of a G4, and often make video production software that isn't compatible with older computers. If you don't want to get *Betamaxed*, or stuck with a dated technology, it might be a good idea to purchase the latest, fastest computer available.

Capture and Storage Systems

Regardless of whether you configure your own system or purchase one ready-made, it's helpful to think about capture and storage before you get started. Decisions you make during the process of importing material to your computer from your tapes shape the outcome of your project. If you employ an organized capture and storage system, you'll have a much easier time finding and keeping track of your footage later on.

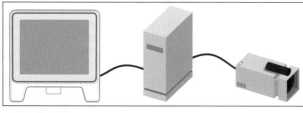
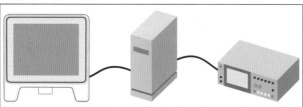

Figure 10-1. A computer connected to a DV camera and a computer connected to a DV deck.

Capture equipment

Capturing DV footage involves playing the material on your deck or camera (Figure 10-1), and then digitally importing the material to your computer via FireWire, also known as an IEEE 1394 connection. It also involves repeatedly fast-forwarding and rewinding through the tape to select the individual shots you want. Video decks were made to fast-forward and rewind without breaking down; cameras were not. Rewinding a tape in your camera can lead to accidentally damaging the tape as it runs through the camcorder at high speeds, and continually fast-forwarding or rewinding can burn out your camera's internal workings (see "The Vulnerability of Any Recorded Media" in Chapter 1). DV decks aren't cheap—they can easily cost more than $1,500—but if using one extends the life of a camera you recently paid $3,000 or $4,000 to purchase, it's a good investment.

SIDEBAR

Configuring Your Own System Versus Buying a Turnkey Setup

When you purchase a new editing system, you have the option of purchasing each component on its own (including editing software, a computer, and associated peripherals such as a deck and displays) or buying a ready-made configuration of hardware and software called a turnkey system. Many video equipment retailers allow you to choose between mixing and matching individual components on your own or opting for a turnkey system they configure for you. Each has its own set of pros and cons.

The advantages of building your own system are:

- You buy only what you want. Since you're configuring the system yourself, you don't have to buy all the elements of someone else's preselected package (which may include things you don't need). You may even be able to use your existing computer, instead of buying a new one.

- You can shop around and contact multiple vendors to find the best price on each individual component. Because you're not buying everything from the same place, you might save money by purchasing different pieces from different retailers who offer competing prices, especially if you buy some online.

- Custom-built computers are modular, and small parts are easy to replace. In contrast, cheaper turnkeys (like some Compaq-HPs, Acers, etc.) have components soldered to the motherboard, so if you need to change a chip or something, you have to buy a whole new motherboard, which is very expensive.

The disadvantages of building your own system are:

- If you're building your system from scratch, it's your responsibility to ensure that everything works. Not everyone likes connecting devices and testing configurations, and if you don't, this probably isn't the route for you to take. If you have trouble assembling the different pieces into a working system, you can't simply call the company who sold it to you for help, because you bought each piece from a different vendor. (A relative once purchased a Sony mini DV camera and Dell computer with video editing software.

He was unable to capture video to his computer from the camera, and at his request, I called the tech support line for both companies. Dell told me they couldn't help me and suggested I call Sony. Sony said they couldn't help me and suggested I call Dell.)

- Searching for the best price and making a series of purchases from multiple vendors takes time and effort. If you're in a hurry to start editing, you may view the time you spend assembling your system as time that you're not spending on editing your film.

The advantages of purchasing a turnkey system are:

- With one purchase, you can bring home a fully functional editing system. The retailer does the work of selecting all the components and delivering them to you as a package deal. This can get you started much more quickly.

- Because you purchase the entire system from one vendor, you have only one merchant to call if there's a problem.

The disadvantages of purchasing a turnkey system are:

- A turnkey system can sometimes cost more than purchasing the individual components from competing retailers (some vendors might charge you a fee for the convenience of buying everything in one easy purchase). I once worked with someone who said everything has an economic value. In this case, the time and effort you save by not having to make multiple purchase decisions may or may not make up for the increase in cost.

- Your turnkey system may include components you don't need and don't use. Because someone else is making the decisions for you, the turnkey system may contain equipment you wouldn't otherwise purchase (it's like the "money saving" option packages on a car).

- They don't always work right out of the box. Your turnkey system may still require varying degrees of assembly and effort before you can get started on your editing project.

Storage hardware

As you set up your capture system, it's helpful to think about your storage system as well, since you'll be using the two together. In recent years, the price of storage has fallen considerably. Independent filmmakers can now buy affordable hard drives that store 120 gigabytes of information or more (which translates to well over 10 hours worth of DV footage) and cost less than $200. Depending on your computer system and how you plan to edit, you can purchase additional internal hard drives that mount inside your computer or external drives that connects to your editing system via FireWire (Figure 10-2). Working with multiple hard drives expands the storage capacity of your computer and helps improve performance. Even if your computer shipped with a very large hard drive, working with an additional drive that you use solely for storing media helps avoid disk *fragmentation* (see the sidebar, "The Problem of a Fragmented Drive") and places less of a strain on the drive running your computer's operating system.

A good external drive should cost about $1 per gigabyte of storage. This means a 120 GB drive should cost in the range of $120. Two good brands are Maxtor (*www.maxtor.com*) and LaCie (*www.lacie.com*). Internal drives should cost slightly less. A web site called dealnews (*http://dealnews.com*) lists the lowest prices on various computer equipment and media, ranging from drives to recordable DVDs. I like this site and their Mac-oriented site dealmac (*http://dealmac.com*), because they don't sell equipment themselves—they just list the lowest prices advertised by other retailers.

Using internal drives

Internal drives often cost less than their external counterparts, and because they reside within your computer, they create less noise. Internal drives can also provide faster *data transfer rates* than an external FireWire drive, which means they can more quickly transfer information from the drive to your computer's processor. For most video applications, good external FireWire drives are fast enough for storage, but I've found that when burning DVDs or encoding material for DVD production, I get better results from media stored on an internal hard drive. For more on DVD production and encoding, see Appendix A.

The main drawback to internal hard drives is they can create substantial amounts of heat. Although desktop model G4 and G5 computers have room to install several internal drives, the documentation for Final Cut Pro advises against installing a total of more than three drives inside a single computer enclosure. According to the documentation, heat build-up can cause problems

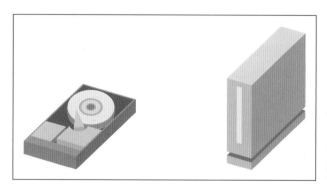

Figure 10-2. Internal and external hard drives.

during capture and playback, and can ultimately lead to the failure of one or more of the drives.

> ── **N O T E** ──
>
> *Working with an internal drive means someone needs to install it. It might be scary the first time you open your computer and start reconnecting components, but it saves you tremendous amounts of money as opposed to bringing it to someone else to do it for you. (Interestingly, many internal drives will work in either a Mac or PC so long as the system is correctly configured. Still, be sure to check that any drive you buy is compatible with your system before you open the box.)*
>
> *Of course, if you aren't computer savvy, or good at putting things together, get someone else to do this. Both for your own safety, and that of your system, don't try to open your computer if you don't know what you're doing.*

Using external drives

External hard drives usually don't have problems with heat build-up, since they're housed outside the computer. External FireWire hard drives are also great for working with laptops, as opposed to extra internal drives, which are generally too large to fit inside a portable computer. Because FireWire drives are *hot-swappable*, meaning they can be connected and disconnected without restarting a computer, they make it easy to organize potentially overwhelming amounts of footage or to separate footage from different projects. You can divide the footage for a large project onto several different drives, and then connect each drive to your computer only when you need it. You can also use the same external drive with more than one computer, which makes editing portable. I have several friends who edit on sophisticated systems in their home studios, and travel with an external drive connected to a laptop so they can edit on the road when time allows or inspiration strikes.

The main downside to external drives is their higher cost, as compared to internal drives, and the fact that you can hear them operating because they're outside your computer's enclosure (if you use a few together, they get really loud). Another potential complication is that external drives generally require a power source to operate, so you have to plug each one into an outlet. This can be an inconvenience if you find yourself running out of available plugs, or if you feel like you already have too many cables and power cords streaming out from your computer.

If you're working with extensive amounts of footage, you might consider setting up a RAID, or redundant array of independent disks. A RAID is a set of disks used to mirror copied data to multiple locations. While these can be very expensive, depending on the capacity of the drives, there are new external FireWire RAIDs that are both reasonably priced and very fast.

The Problem of a Fragmented Drive

Hard drives store files in chunks of information. When there's room on a disk, a computer can place large chunks of information close together, which makes them easy to read. When the drive starts to run out of room, the computer places smaller chunks of information wherever it can find space on the disk. If parts of a file are scattered throughout a disk, instead of stored near one another, the file becomes harder to read and takes longer for the computer to process. When this happens, the disk has become fragmented and the computer's performance suffers. Video files are notorious for fragmenting hard drives, especially as you delete older files and record new ones.

Two software applications you can use to help maintain and defragment your drives are Diskwarrior by Alsoft (*www.alsoft. com/DiskWarrior*), and Norton Systemworks by Symantec (*www.symantec.com/sabu/ sysworks/basic*). DiskWarrior runs on Mac; Systemworks is now a PC-only product, although earlier versions were cross platform.

If you're running Windows XP on your computer, the operating system includes software called Disk Defragmenter, which can be scheduled to perform periodic maintenance on your PC.

NOTE

Make sure you plug the correct power supply into your drive. Most power supplies look similar, but if you plug into one that has a higher output voltage than your drive takes, it will fry the drive. To avoid confusion, you can color-code your drives and power supplies with colored tape, or even label them.

If you're working on a project that is your life, you might consider getting an extra drive, periodically backing up the data and project files, and keeping it at a friend's place, in case something happens to your house (fire, flood, earthquake, etc.).

If you have more than one external drive, especially if they are all the same model, you can label each with the name of a project, or a project name and a number, to avoid confusion.

Why removable media is not an effective choice

Instead of purchasing an external hard drive, it may be tempting to work directly from footage stored on removable media, such as a Zip disk, a data DVD, or a CD. This is a bad idea. A computer won't be able to read the information off these discs fast enough to play the video you store on them. You can archive media onto a Zip disk, data DVD, or CD, and then copy the files to your hard drive to work with, but otherwise, removable media just won't work. To meet the needs of a video editor, any storage device needs to transfer information to and from a computer at a consistently high rate. Final Cut's documentation recommends using drives, either external or internal, with sustained data transfer rates of 8 MB per second or more (this recommendation also applies if you're running another editing program, such as Avid Xpress or Adobe Premiere). The documentation also recommends using a disk that spins at 5,400 *revolutions per minute* or higher (many editors recommend an even higher 7,200 rpm) because the faster a disk spins, the more quickly it can transfer information and access individual files on-demand. CDs and Zip disks don't make the cut because they just aren't fast enough.

Keeping your footage organized

Once you've got your capture and storage system set up, it's important to establish and maintain a tape numbering system. The inexpensive nature of DV footage makes it easy to shoot hours and hours of footage, so before you know it, you may have a shelf full of tapes.

Professional cameras enable a filmmaker to manually set the start timecode that appears on a tape. Filmmakers often use this to organize their footage. For example, a filmmaker might set the start timecode of the first tape to 01:00:00;00, the second to 02:00:00;00, the third to 03:00:00;00, and so on. As a result, each tape will have a different source timecode. This enables a filmmaker to set different timecode on up to 24 different tapes (after 23:59:59;29 the timecode recycles to 00:00:00;00).

Unfortunately, most prosumer cameras don't offer a function like this, so each tape will have timecode that's identical to the others. This is a potential problem once the footage has been imported to the computer, because if two clips share the exact same timecode, it can be hard to tell them apart in later stages of postproduction.

An easy solution to this problem is to number each tape (place the same number on the tape label and the box) and use the same number to identify the tape as you capture the footage to your editing system. Project and pointer files store not only the timecode information for each edit you make, but the tape source of all the material in your film. This source information is especially important if you need to recapture your footage in the event of a system crash, or if you decide to output an EDL and bring your project to a high-end post facility to do something beyond the ability of your system (see Appendix A).

The Importance of a Well-Calibrated NTSC Monitor

When you're editing digital video, the images you see on your computer screen won't match what your audience sees on a television. NTSC monitors and computer screens display colors and motion differently, and they also display different areas of a video frame.

NTSC monitors display a more limited range of color than computer screens. As a result, colors that look really good on a computer screen might not look that great to audiences watching your work on a television. Even if you never expect a television broadcast, the way

Figure 10-3. Photo of a well-labeled DV tape.

> **NOTE**
>
> *A good DV tape label (Figure 10-3) not only contains a reel number but also notes the date of the shoot, the name of a project, and a very brief description of the tape's contents. When you start working with large amounts of material, it's easy to lose track of what footage is really on which tape. Taking careful notes and effectively using your tape labels always pays off.*

SIDEBAR

Developing a Tape Naming Convention

A naming convention is a set of rules you apply to ensure the names you give tapes or files in a project all follow the same logic.

A very straightforward and widely used convention involves numbering each tape sequentially, starting with 001. The next tape would be 002, then 003, etc. This enables a filmmaker to use 999 separate tapes without repeating a tape name. (I don't know anyone who records that many tapes, but it's nice to have the option.)

Some filmmakers also add the letters "CR" before the number, so their tapes are labeled CR001, CR002, CR003, etc. CR stands for camera reel, and is a holdover from the days when people shot reels of film.

Regardless of the naming system you come up with, the key is to implement a numbering system at the very start of your project and stick with it through the very end of postproduction. During the capture process, Final Cut Pro and Avid both ask editors to enter tape names whenever they begin to capture media, or whenever they change cassettes during the capture process. Entering exactly the same name that's written on a tape's label offers an easy way keep a tape numbering system intact, and helps avoid creating duplicate or multiple tape names. At first, entering the wrong name for a tape might not seem like much of a problem, but it can ultimately make it impossible to find the original source of a clip. Because many editing systems are case sensitive, CR001 would be considered a different name than cr001. Entering an extra space could also cause problems, for example CR 001 and CR001 would be considered different tapes. All this may seem like a lot to keep in mind, but like anything else in the world of film, a little organization and planning at the start of a project can help prevent tremendous headaches and hair loss later on.

NOTE

There's just no substitute for a dedicated NTSC monitor, especially when you're working with composites and motion effects. If you're editing video that audiences will watch on a television or video monitor, adding an NTSC monitor to your system is the only way to know what they'll really see.

your work looks on television is still important. If you plan on distributing your work on DVD or VHS, the people you give copies to will almost certainly be watching your film on a television screen. If the colors don't look the way you want them to, showing people your work may turn into a very disappointing experience. (For more information on the difference between video monitors and computer screens, see Chapter 11.)

NTSC monitors and computer screens operate at different scan rates, so the images onscreen refresh more frequently on a computer monitor than on a television. This becomes particularly noticeable when you start to work with motion control effects during editing, for example, speeding up your footage or slowing it down. Digital editing systems enable filmmakers to create complex motion control sequences where images move across the screen in dramatic slow motion, or at fantastically high speeds, to create some very cool effects. The only way to judge the effectiveness of a motion control sequence is by using an NTSC monitor (Figure 10-4). Because the images on a computer screen refresh more quickly than the images on an NTSC monitor, the motion of an effects-heavy sequence may look super smooth on a computer, but appear choppy or jerky when output to video. If you preview the results on an NTSC monitor, you can spot potential problems long before you deliver your work to a client or a film festival screening committee. If you rely solely on the image on your computer screen, you're taking a big chance.

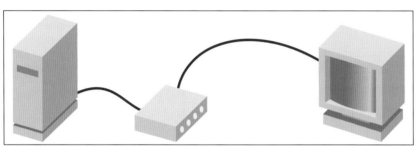

Figure 10-4. An NTSC monitor connected to a computer via a bidirectional converter.

Lastly, computer monitors display a greater area of the frame than you'll see on an NTSC monitor. If you look at a frame of video on a computer screen, and look at the same frame on an NTSC monitor, you'll notice the NTSC monitor cuts off the edges on all four sides of the frame. This makes viewing your work on an NTSC monitor especially important when you're creating titles. Text you place on the screen may look great when you view your work on a computer, but on a television screen, large chunks may get cut off because NTSC monitors display a different part of the frame. If you're taking the time to make a really sharp looking title, it's worth the added step of viewing your work on a dedicated monitor to ensure the audience sees what you want.

Connecting a monitor to your system

You can easily connect your editing system to an NTSC monitor by routing the digital output of your computer through a camera, or through a media converter. Using a camcorder or DV deck is the easiest way—you simply connect the computer to your camera via FireWire, and connect the camera's RCA video and audio outputs to your monitor. In order for this to work, your camera needs to be connected and turned on. If you don't want to use a camera, you can purchase a bidirectional media converter, which converts video signals from analog to digital, and from digital to analog. The converter's analog-to-digital capabilities also enable you to capture analog video footage to your computer, such as older home movies shot in hi-8. The digital-to-analog conversion allows you to output your work to an analog video deck or an NTSC monitor. Affordable options for a bidirectional converter include the Canopus ADVC-110 (Figure 10-5), which sells for less than $300 (*www.canopus.com/US/products/advc-110/pm_advc-110.asp*). Canopus also sells a PC-based prosumer editing package complete with a

media converter, editing software, and a video capture card designed to expands the video processing abilities of your computer. The package sells for about $700 (*www.canopus. us/US/products/EDIUS_DV_Pack/ pt_EDIUS_DVpack.asp*).

More expensive options include the Io series of external video capture devices from AJA (*http:// www.aja.com/ products_Io.html*), which retail for about $1,200. Both options provide good quality output, AJA's Io series, also shown in Figure 10-5, offers a variety of professional audio and video outputs, enabling you to connect your editing system to a wider array of high-end audio and video decks.

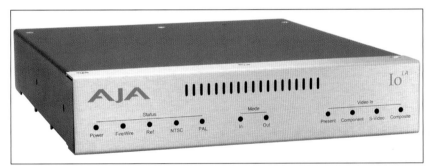

Figure 10-5. The Canopus ADVC-110 and the AJA Io LA External Video Capture Device.

Regardless of the setup you decide upon, putting together your own editing facility provides a level of freedom and accessibility that filmmakers could only dream about a few years ago. The cost of a new computer and the video equipment that goes into assembling an editing system is certainly nothing to sneeze at, but if you had to rent time at another facility to finish your project, you could easily pay more just to make one film. For years editors yearned for the ability to edit on a budget, without getting a budget-looking product. Now, thanks to DV, you don't have to spend a fortune to make a film that looks like a million bucks.

Chapter 11 explores image creation and transparency, and ways you can incorporate graphics into your digital video projects. Your editing system not only provides you with the ability to cut shots together and build a story, but to add layers of polished effects to your work—whether you prefer a subtle visual aesthetic, something totally over the top, or a balance between the two.

SIDEBAR

Connecting a Monitor: Step by Step

You can use a media converter to connect your desktop or laptop editing system to the monitor of your choice. Although the process might seem daunting if you've never done it before, it's really straightforward:

1. Connect a FireWire cable to a compatible FireWire port on your computer (if the plug fits, it's compatible).
2. Connect the other end of the cable to your media converter.
3. Connect the video output of your media converter to the input on your monitor.

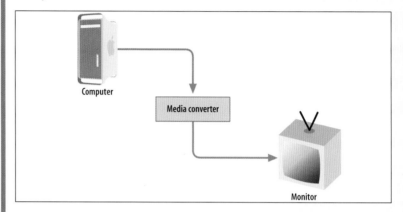

Different NTSC monitors use different types of connections. Professional-quality monitors use component video, which essentially breaks an RGB video signal into three separate signals, one for each color. Prosumer models generally use composite video, which is a single video signal that uses a composite of all three component video colors. Check your monitor's documentation to see what type of input it takes and what type of cables you'll need.

An Overview of Composite Images

11

When you edit together a group of shots, they form a sequence. A *sequence* is the basic building block of any film—an individual shot on its own may be very powerful, but a sequence begins to tell a story. (When people talk about a "scene" in a film, in cinematic terminology, they're talking about a sequence.) By stringing together a group of sequences, a filmmaker creates an *arc*, a story component that builds steadily to a climax and then resolves to a clear ending. As an editor, you never know how two shots will combine until you actually watch them together onscreen. Working in the Timeline of your digital editing system, you can place single shots one after another and build a basic sequence, or you can combine shots on multiple layers and begin to build a *composite*.

A composite video effect combines multiple images into a single product. The term *composite* can be used as a noun (as in, "This sequence is a composite of two video clips: a still image and a computer-generated graphic") or as a verb (for example, "It took me all day to composite the actor into the background").

When you composite video, you have complete control over the placement and opacity of each video clip, still image, and graphic that appears in your sequence. You can determine which areas overlap, which parts of an image are completely transparent, and which images obscure parts of the frame behind them. With a combination of technical skill and your creativity as a filmmaker, you can construct intricately layered visual compositions that provide the perfect complement to your story.

This chapter walks you through the process of importing material into Final Cut Pro and After Effects, editing the material into your Timeline, and then building and adjusting a composite sequence. First, the chapter explores transparency and color in digital video and explains how they make composite sequences possible.

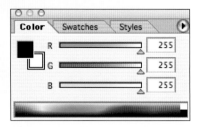

Figure 11-1. The Color palette in Photoshop displays, and enables you to adjust, the red, green, and blue values of any color. In this figure, the RGB sliders are set to white.

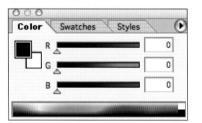

Figure 11-2. In this figure, the RGB sliders are set to black.

The Importance of RGB

All screen images, meaning anything you create for display on a computer or video monitor, are RGB images. Print images, however, use a different color process called CMYK (cyan, magenta, yellow, and black). While RGB is additive (if you add all the colors together you get white) CMYK is subtractive (if you add all the colors together you get black).

To make sure your image looks the way you want it to, and to avoid problems when you import it to video, always work in the RGB color space.

What an Alpha Channel Does

Although computers and televisions process images differently, they both use the same three colors to create images: red, green, and blue. Using an *additive* color process, computers and televisions combine red, green, and blue (RGB) at varying levels of intensity to create the full spectrum of colors you see onscreen. Because the process is additive, combining all three colors at their fullest intensity creates white (as in Figure 11-1). When all three colors display at their lowest intensity, the result is black (Figure 11-2).

Professional strength image creation programs, such as Photoshop, and many digital video programs, such as Final Cut Pro and After Effects, assign each color to its own *channel*. As a result, RGB images contain three channels of color information, which can be manipulated separately, or in combination, to produce various effects. RGB images also contain a fourth channel, called an *alpha channel*, which controls transparency. Working with the alpha channel of an image enables you to create large or small areas of transparency, which you can use to create a wide range of effects.

Final Cut Pro and other editing applications allow you to create an image in Photoshop and then import the image, including its transparent areas and alpha channel information, into a sequence of video clips. (After Effects is especially Photoshop friendly—both programs are manufactured by Adobe, and they're engineered to work together.) By combining multiple images and cleverly manipulating their alpha channels you could, for example, make it look like two people were standing next to each other even though they were photographed in separate locations. If you're not interested in playing tricks on your audience, you can use alpha channels to create title sequences or onscreen text to introduce interview subjects and identify locations, then superimpose the text and graphics over full motion video.

In the 2002 documentary *The Kid Stays in the Picture*, the directors layered multiple still images and edited the alpha channels to create both depth and motion. Combined with voiceover narration about the career of actor and movie producer Robert Evans (the documentary explored his life in Hollywood), these image sequences added a stunning visual impact and power to the film. If the filmmakers had used a more conventional approach—for example, pairing still images with people speaking into the camera—they still would have had an interesting story, but the use of inventively constructed montage sequences made the film much more interesting to look at.

Importing Images and Adding Them to the Timeline

Before you can work with video footage or still images in either Final Cut Pro or After Effects, you first need to import them into a project. When you import video footage or a still image into either program, your computer doesn't make another copy of the material. Instead, both programs create links to the material's location on your hard drive. This saves disk space and enables you to edit without destroying your original footage (see Chapter 9). Once you've imported material into your project, you can add it to your project's Timeline and start cutting a sequence together.

Importing to Final Cut Pro

To import footage to Final Cut Pro, follow these steps:

1. Open Final Cut Pro. By default, Final Cut Pro opens the most recent project created on your computer. You have the option of working with the project that opens when you start the program or working with another project.

To open an existing project, select File Open and navigate to the project you'd like to work with.

To create a new project, select File New.

2. Select File→Import→Files. The Choose a File window opens.

3. Select the file or files you want (you can select multiple files by holding down the Command key as you click).

4. Click Choose to close the window and import the files. Final Cut imports the material into your project.

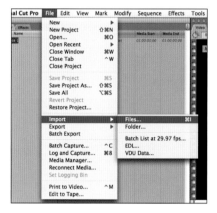

----- NOTE -----

To recreate the Final Cut Pro sequences illustrated in this chapter:

1. Navigate to the Ch11 folder in the Exercises folder on the StartToFinish DVD that came with this book. Drag the folder to your hard drive (don't try to work with video files directly from a CD—see Chapter 10).

2. Import stopSignEdit.psd, which contains an image of street signs on a transparent background, and background.mov, which contains an animated tangerine-colored background.

3. Follow the step by step instructions in this section and those that follow.

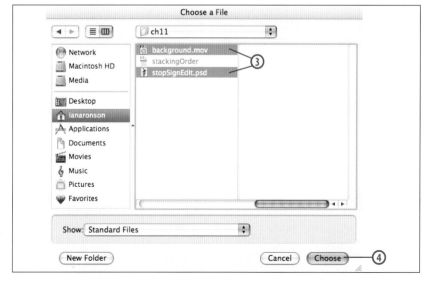

Creating a Final Cut Pro composite

1. Select a clip of imported material in the Browser window, and drag it into a video track on the Timeline. The Browser lists all the material available in your project, and the Timeline is where you arrange shots. (For more details on Timelines and frames, see Chapter 9.)

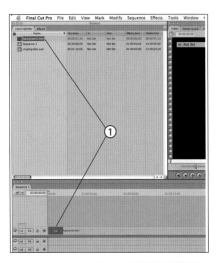

2. Add another clip to a higher-level video track in the Timeline. The easiest way to do this is to drag a clip from the Browser to a location just above a clip in your current Timeline. When you release the mouse, Final Cut Pro adds a new track to the Timeline to accommodate the clip you're adding. This creates a basic composite image—if the image on your top-layer track contains areas of transparency, the image on the bottom-layer track will show through.

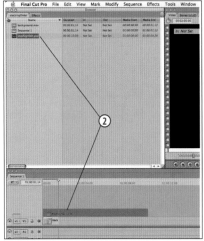

3. Preview the composite in the Canvas window. You won't be able to play the composite in the Timeline until you render, but you can preview your composite using the arrow keys on your keyboard. To scroll through the sequence and simulate playback, bring the Playhead back to the start of the sequence and use the arrow keys on your keyboard to move through the sequence frame by frame.

4. Save your work.

Importing to After Effects

To import footage to After Effects, follow these steps:

1. Open After Effects. Unlike Final Cut Pro, After Effects does not automatically open the last project you worked on. As a result, you can either work in the new project that After Effects creates when you open the program, or you can open an existing project.

   To open an existing project, select File→Open Project and navigate to the project you'd like to work with.

2. Select File→Import→File.... The Import File window opens.

3. Select the file or files you want. You can select multiple files by holding down the Command key (Mac) or the Ctrl key (Windows) as you click.

   Click Open to import the files and close the window. After Effects imports the material into your project. Depending on the files you import, the Interpret Footage window may open. For more on this window, see the sidebar "Interpreting Channels and Layers."

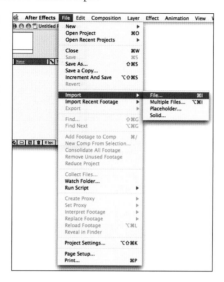

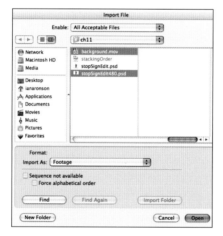

Interpreting Channels and Layers

Unlike Final Cut Pro, After Effects enables you to specify how it handles images with alpha channels or multiple layers.

Not all image editing applications label their alpha channel types. Photoshop labels the alpha channel in an image as either straight or pre-multiplied (for a detailed explanation of these terms, see "Alphas, composite modes, and blending modes" later in this chapter). If After Effects cannot determine whether an image has a straight or premultiplied alpha channel, the Interpret Footage window opens.

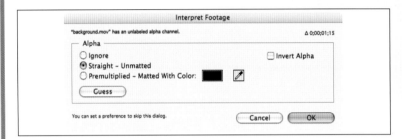

The Interpret Footage window gives you the following options:

Ignore This option disregards any transparency information in an image, so it appears fully opaque.

Straight After Effects interprets the image as containing a straight alpha channel.

Premultiplied After Effects interprets the image as containing a premultiplied alpha channel.

Guess After Effects determines the alpha type without your help.

When you import a layered Photoshop image, After Effects opens a dialog box with the name of the file you're importing. (In the illustration, I was importing a file named stopSignEdit480.psd, so the name of the dialog box became SignEdit480.psd.)

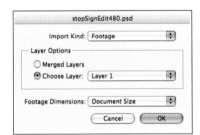

The Import Kind pull-down menu enables you to treat the file as a Composition, or as Footage. A Composition in After Effects is similar to a Sequence in Final Cut Pro or Premiere—an editor places clips in a Composition and arranges them as he would clips in a sequence.

Choosing Composition imports the file into your project with its layers intact.

Choosing Footage imports the file as a clip, which you can then add to a Composition. When you choose Footage, the Layer Options radio buttons become active and you can choose Merged Layers or Choose Layer. Merged Layers imports all the layers. Selecting a layer of the Photoshop document from the Choose Layer pull-down menu imports that layer only.

Creating an After Effects composite

Unlike Final Cut Pro, which opens a new sequence when you create a new project, After Effects doesn't automatically create a new composition. So, if you're working with a new project, the first step is to manually open a new composition.

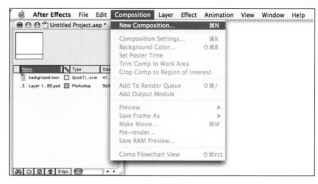

1. Select Composition→New Composition. The Composition Settings window opens.

2. From the Preset pull-down menu, select NTSC DV, 720×480. (If you're using a format other than mini DV or DVCAM, select the appropriate frame size. For more detail, see the sidebar "Video Formats and Frame Size" later in this chapter.)

3. Enter 0:00:10:00 in the Duration field. This sets the duration of the composition to 10 seconds, which is the same length as the *background.mov* clip.

4. Click OK to close the dialog box. The Composition window and Timeline open.

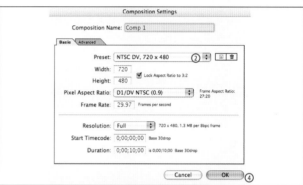

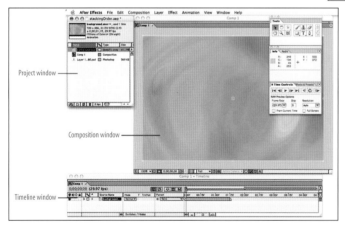

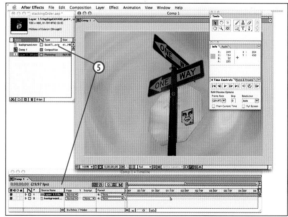

5. Drag a clip from the Project window into the Timeline and release the mouse button. The clip appears in the Timeline and displays in the Composition window.

6. Add another clip to a higher layer in the Timeline. You can do this by dragging a clip from the Project to a location just above a clip already in the Timeline. When you release the mouse button, After Effects creates a new layer that contains your newly added clip. This creates a basic composite image—if the image on your top-layer track contains areas of transparency, the image on the bottom-layer track will show through.

7. Save your work.

Figure 11-3. An unedited image of street signs, with its background intact.

Figure 11-4. An edited version of the image in Figure 11-3 with its background removed.

── **NOTE** ──

For more information on straight versus premultiplied alpha channels, visit http://www.adobe.com/ support/techdocs/317477.html.

Alphas, composite modes, and blending modes

As explained earlier in this chapter, an alpha channel determines which parts of an image become transparent and which remain visible to an audience. Figure 11-3 shows a still image of a street corner near my apartment. Using Photoshop, I removed the background, leaving only the street signs surrounded by an area of transparency (Figure 11-4). As I removed the background, I was also modifying the image's alpha channel.

The alpha channel itself is not visible to a film viewer, but editing applications will often display an image's alpha channel while you're working (viewing a clip's alpha channel can help you manage transparency). Figure 11-5 shows a screen capture from Final Cut Pro that displays the alpha channel of the edited image. The black areas in the screen capture display as completely transparent. The white areas remain opaque. This image uses a straight alpha channel, which means sections of the image are either 100 percent transparent or 100 percent opaque. Some images that contain translucent visuals—for example, lens flare—use a *premultiplied* alpha channel. A premultiplied alpha channel image enables background layers to show through translucent areas, while also containing sections that are either fully transparent or fully opaque.

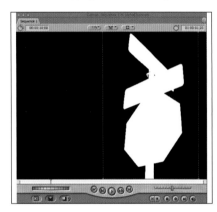

Figure 11-5. The image's alpha channel.

Experimenting with stacking order

Final Cut Pro and After Effects both use layers to organize, or stack, images one on top of another. The *stacking order* of images determines which images will be visible over the rest. Final Cut and After Effects both use the same stacking order as Photoshop: images on higher levels obscure images on layers underneath. I often explain this by comparing the stacking of layers in one of these programs to the visibility of papers on a messy desk—depending on how the papers are arranged, the top sheet either partially or fully obscures the papers below. Likewise, the opaque areas of a higher-layer image hide what appears behind them, while transparent areas allow a viewer to see through to lower layers of video.

Depending on the level at which you place an image, and the ways you use an alpha channel, you can make objects pass behind or in front of it. I once took a still image of a photographer, removed the background, then placed the image in a sequence to appear behind the end credits. I accidentally placed the still on a higher video track than the text, so the image of the photographer appeared in front of the credits, partially obscuring them. As

I was about to move the still image to a lower video track, I realized I had stumbled into a cool effect by removing the background. When I previewed the sequence, the credits scrolled up from behind the photographer, and became fully readable as they moved toward the top of the screen. I decided to keep the image and its alpha channel on the top video track, and when I showed the sequence to the client, he really liked it.

I created a new sequence to demonstrate the effect, which is pictured in Figure 11-6 through Figure 11-8. The sequence contains three video tracks (Figure 11-9). The bottom-level, or base-level track, is the background image. The top-level track is the stop sign image. The text "Stacking order" appears in between. As the text scrolls up from the bottom of the screen, it's partially obscured by the stop sign image. The transparent areas of the stop sign image on the top layer allow the text to show through, the opaque areas of the image block out anything on the two layers underneath.

Figure 11-6, 11-7, 11-8. The words "Stacking order" appear on a layer underneath the street signs image. Because of the stacking order of the video tracks, parts of the text are obscured by the signs.

Figure 11-7.

Figure 11-8.

The *Exercises* folder on the DVD contains After Effects and Final Cut Pro project files that demonstrate how this sequence is constructed. To access these files, insert the DVD in your computer, open the *Exercises* folder, then navigate to *ch11→stackingOrder.aep* (After Effects) or *stackingOrder.fcp* (Final Cut Pro). You need to have After Effects and/or Final Cut Pro installed on your computer to open these files.

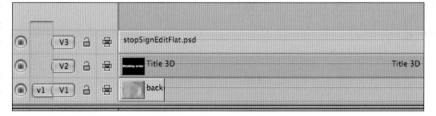

Figure 11-9. This screen capture of the Final Cut Pro Timeline illustrates the stacking order in the sequence pictured earlier in this section.

Rearranging stacking order

In After Effects or Photoshop, you can easily rearrange the stacking order of your layers by simply dragging a layer to a new position in the Timeline. When you release your mouse, the new stacking order takes effect.

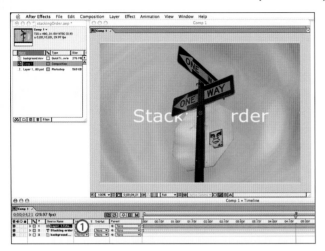

You can follow along with this exercise using a project file you created earlier, or by opening the After Effects project file *stackingOrder.aep* in the *ch11* folder on the DVD. (Remember to drag the folder into your hard drive before opening it.)

1. Click the top layer image in the Timeline to select it.

2. Drag the image one layer down, between Stacking order and *background.mov*, then release the mouse. After Effects updates the order of your layers, and the text now appears in front of the signs in the Composition window.

3. Drag the background image, *background.mov*, one layer up and release the mouse. After Effects updates the stacking order. The text on layer 1 is still visible as the top-layer image. Because the top layer contains a large area of transparency, the orange background clip on layer 2 remains visible as well. However, because the image on layer 2 is completely opaque, the stop sign image underneath is now completely obscured.

Composite modes and blending modes

In addition to using alpha channels, Final Cut Pro employs a variety of composite modes, which determine how a layer of images will interact with the layer beneath. After Effects does the same thing using blending modes. The default mode for both applications is Normal. In Normal mode (Figure 11-10), the image on the highest video track completely obscures any images on lower tracks. In the case of an image on a transparent background, the visible parts of the image obscure any material that appears behind them on a lower track. Working in Normal mode, you can overlap different images, video streams, and text to create complex multilayered composites.

Instead of using normal mode to obscure images, you can use one of the many other composite or blending modes to blend images together. Final Cut analyzes the color values in two layered clips and then uses different mathematical formulas, depending on the mode you select, to combine their images in a variety of ways. In Subtract mode (Figure 11-11), Final Cut subtracts the mathematical values of the colors in the street signs image from the color values in the background image below. The result is a darker composite of the top-level clip, while the background remains unchanged. In Hard Light mode (Figure 11-12), the lighter areas of the top-layer image become transparent (the idea is that hard light simulates shining a bright spotlight on the top layer). In Figure 11-12, the background shows through the white areas of the one-way signs, and the back of the stop sign. Final Cut and After Effects both offer long lists of composite modes, if you're not sure which makes the best choice for your footage, just put two clips together in your Timeline and experiment.

Figure 11-10. A screen capture of a layered sequence in Final Cut Pro in Normal mode

Figure 11-11. The sequence from Figure 11-10, shown in Subtract mode.

Figure 11-12. The sequence from Figure 11-10, shown in Hard Light mode.

To change the composite mode of a clip in Final Cut Pro

To change the composite mode of a clip in Final Cut Pro, do the following:

1. In the Timeline, click on the clip you would like to change. This will select it and make it the active clip. The composite mode of a clip determines how it will interact with the clip on the layer underneath.

2. Select Modify→Composite Mode. A new submenu opens.

3. Select a composite mode from the list. Final Cut Pro will apply the composite mode to the clip. You can then decide if you like the change or might prefer another.

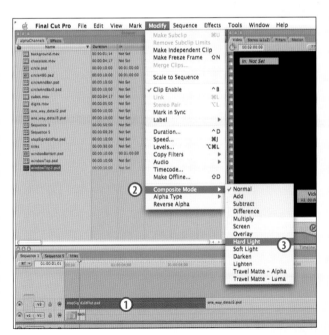

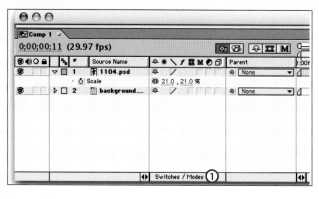

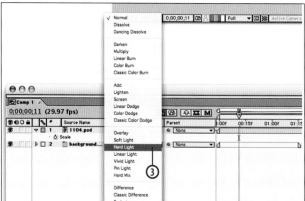

To change the blending mode of a clip in After Effects

To change the blending mode of a clip in After Effects, do the following:

1. Click "Switches/Modes" at the bottom of the Timeline window. The Mode pull-down menu appears.

2. Click the Mode pull-down menu of the clip you'd like to change. A long list of blending modes opens. Blending modes in After Effects determine how a layer of images interacts with the layer beneath it. This composition has two layers, so I'm changing the blending mode of the top-layer clip.

3. Select a mode from the list. After Effects applies the mode you choose.

Rendering Your Work

Whenever you create a composite sequence, as you did earlier in this chapter, or whenever you add a transition or execute an effect, you generally need to render the sequence before you can play it back. Rendering a sequence creates a new series of video clips that contain your source footage and any composites, transitions, or effects you've added. As explained in Chapter 10, your computer's rendering speed depends on the processor and available RAM, as well as the nature of the sequence itself (the more effects, the more time it will take to render, and some effects take longer than others). Depending on your computer, you may be able to preview some effects without rendering, but even the most muscular computer system requires render time before you can view intricate composite effects. Longer and more complicated sequences or compositions take more time to render, and if you have any other applications running on your computer (for example, if you like to listen to iTunes while you work), they'll divert power from your processor so rendering will take even longer. (To avoid problems, it's a good idea not to do anything else with your computer while you render.)

Rendering a longer project, such as an effects-intensive feature film, can easily take more than 24 hours. Rendering the short sequences you created in this chapter will probably take a few minutes or less.

Because rendering creates new video files, it's important to make sure you have enough hard drive space available before you start. If you run out of space part way through, your computer won't be able to complete the rendering process. (For more information on storage systems and drive space, see Chapter 10.)

NOTE

Some editors like to render as they go along, meaning they render their work each time they try a new effect or make a change to an existing one. Other editors like to wait until they've added a few effects to a sequence and render them all at once.

Rendering as you go enables you to see all the details in your effects and transitions as you add them to your project, but it also means you have to stop working while you wait for each individual item to render. (Some people like this because it enables them to take breaks from editing.)

Rendering all the elements in a sequence or composition at once consolidates your waiting time—it doesn't shorten the overall time required for your computer to render the material, but it enables you to continue working until you finish a preliminary draft.

SIDEBAR

The Problem of Overcompressed Media Files

When you output your work from either Final Cut Pro or After Effects, you have the option of choosing various compression settings. The idea behind video compression is that video files require substantial amounts of memory to store, and considerable amounts of processor power to play on a computer. When you compress a video file, the application you're using removes information to make the file more manageable and, as a result, easier to store and to play back. (For an overview of the compression process, see Chapter 3.)

Compression reduces the number of bits/bytes in a file. In theory, compression applications remove redundant information that viewers don't notice (the idea is that uncompressed video files contain more visual information than audiences can mentally process). In practice, however, compressing a file often leads to a loss in quality. When material is overcompressed, visual imperfections called compression artifacts appear in the footage. These generally take the form of blocky patterns that appear around areas of fine detail. (If you've ever tried to compress a still image as a .jpg for use on a web site, you may have noticed similar problems with artifacts and pixelation—parts of an overcompressed image can start to look like a mosaic.)

When you're outputting a master version of your project, and especially when you're outputting material with titles and graphics, avoiding problems with compression becomes particularly important. Even small artifacts in your movie will distract the audience from your story and take away from the quality of your work (even if viewers don't consciously notice artifacts, they'll realize that something doesn't look right).

To avoid problems with compressed footage, output the master version of your file at the highest possible quality. This chapter contains step-by-step exercises that show you how to output your work as uncompressed QuickTime files from both Final Cut Pro and After Effects. You can then bring your uncompressed file to a postproduction facility or use a specialized compression application, such as Apple's Compressor or Discreet cleaner, which is available in PC and Mac versions, to create files compressed for specific purposes. (Appendix A of this book describes the process of bringing your project to a professional facility to create an HD master or film print, and it also describes the process of compressing media yourself to create a DVD or to stream your work from a web site.)

For more information on compression settings and output quality in After Effects, see "Choosing compression options" in the After Effects Help files.

For more information on compression settings and output quality in Final Cut Pro, see "Exporting a QuickTime File" in the Final Cut Pro Help files.

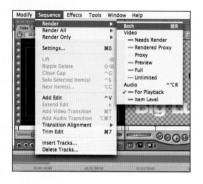

To render a sequence in Final Cut Pro

To render a sequence in Final Cut Pro, do the following:

1. Select Sequence→Render→Both. Final Cut Pro renders any unrendered material in your sequence (this includes both video and audio).

Final Cut opens a new window, displaying the percentage of the material in a sequence that has been rendered so far, and the estimated time remaining in the render process.

When rendering is complete, Final Cut Pro closes the Writing Video window, and your sequence is ready to play back.

2. Play your newly rendered sequence, and enjoy the satisfaction of watching your effects playback in real time. (Much nicer than dragging the Playhead through a sequence to preview an effect, isn't it?)

3. Save your work.

Setting the location of your render files in Final Cut

Final Cut Pro enables you to select the location where your render files will be stored, which is called the scratch disk. When you render a file in Final Cut Pro, the application generates a new set of video files containing the effects you've added. Setting a new scratch disk location can be very helpful if you have more than one hard drive installed on your system and one starts to fill up. Because Final Cut Pro stores your project files and media files separately (render files are treated as media), you can store different types of files on separate hard drives. (For more on the difference between project files and media files, see Chapter 9.)

To set render the scratch disk:

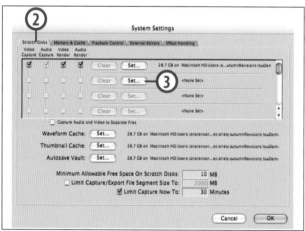

1. Select Final Cut Pro→System Settings.... The System Settings window opens. In the illustration, the Video Capture, Audio Capture, Video Render, and Audio Render are all set to the same location on the main hard drive of my computer.

2. Click the Scratch Disks tab to access the Scratch Disks controls.

3. Click the Set button in the second row. The Choose a Folder window opens.

4. Navigate to the folder you'd like to use for your Video Render and Audio Render scratch disks.

5. Click Choose. Final Cut Pro chooses the folder you've selected as a new scratch disk and closes the Choose a Folder window. The System Settings window remains open. The next step is to designate the new scratch disk as the destination for newly rendered audio and video files.

6. Click the checkboxes in the second row under Video Render and Audio Render. This designates the new scratch disk you created in step four as the destination for your video and audio render files.

7. Uncheck the boxes in the first row under Video Render and Audio Render. Doing this deselects your original scratch disk.

8. Click OK. Final Cut Pro finalizes the changes to your scratch disk settings and closes the System Settings window.

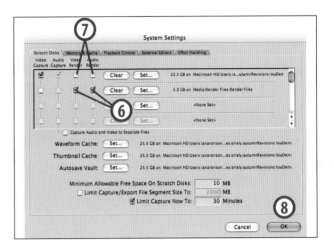

To output an uncompressed QuickTime file from Final Cut Pro

If you want to output a high-quality video file directly from your editing system, you can choose to export an uncompressed QuickTime file. Depending on your plans and your budget, you can also export an edit decision list, or EDL. You can then bring your EDL to a high-end postproduction facility to create a broadcast-quality version of your project with the help of the facility's staff. Because you're working with Final Cut Pro, you can also use a Mac-only application called Compressor to output a compressed version of your project to distribute as a DVD or to stream on a web site. (Creating an EDL and compressing a video file are both covered in Appendix A.)

As mentioned in the sidebar "The Problem of Overcompressed Media Files," exporting an uncompressed QuickTime file provides you with flexibility. An uncompressed media file is the highest quality video file you can output directly from your computer without using any other equipment. Exporting at the highest quality possible gives you the option of compressing a file in the future—if you export at a lower quality, you don't have the option of increasing the quality level later on.

Now that you've rendered your sequence, you can also output your work as a discrete video file. This exercise shows you how.

1. Select File→Export→QuickTime Movie.... The Save window opens.

2. Navigate to a location on your computer's hard drive where you'd like to save the file. In this case, I'm exporting the file to a folder called *rendered files*.

3. Enter a name for your file in the Save As field. In the example, I named the file *sequence1.mov*. Adding the filename extension *.mov* is very important, it identifies the file as a QuickTime movie. If your filename doesn't contain the *.mov* extension, you may have trouble playing it back later—especially if you're working on a Mac and giving files to someone who uses a PC.

4. From the Setting pull-down menu, select Uncompressed 10-bit NTSC 48 kHz. This setting creates an uncompressed QuickTime file. Because the file is uncompressed, the images appear at a higher quality. (For more on compressed versus uncompressed media, see "The Problem of Overcompressed Media Files" earlier in this chapter.)

5. Click Save. Final Cut Pro closes the window and generates a high-quality video file from your sequence. You can now bring this file to a postproduction facility to strike a film print or create a broadcast-quality HD master. You can also import the file into a compression application to create a DVD. Whatever you decide to do, creating an uncompressed video file provides you with options. You can downsample a file to create a lower resolution video file in the future, but you can't do the opposite.

Rendering a composition in After Effects

As with many other elements of postproduction, rendering a composition in After Effects is a very different process than rendering a sequence in Final Cut Pro. When you render an After Effects composition, the application generates and automatically *exports* a new video file.

This exercise shows you how to export your composition as an uncompressed QuickTime video clip that you can then import into a digital editing application, such as Adobe Premiere or even Final Cut Pro. (For a full list of supported file output types, see "About rendering" in the After Effects Help files). You can also re-import the rendered video file into After Effects to create additional composite effects.

SIDEBAR

Setting the Work Area

You can render a select section of your After Effects composition by defining the beginning and end of the work area. The work area is essentially the active part of your Timeline. By marking in and out points, you can render a specific area of the Timeline and ignore others.

To mark the start, or in point, of the work area:

1. Move the Current Time Indicator to the point in the Timeline where you'd like the work area to begin.

2. Press the letter B on your keyboard.

To mark the end, or outpoint, of the work area:

3. Move the Current Time Indicator to the point in the Timeline where you'd like the work area to end.

4. Press the letter N on your keyboard. Markers now indicate the beginning and end of the work area in your Timeline.

This method works for both PC and Mac versions.

To render a composition:

1. Select Composition→Make Movie.... The Render Queue window opens.

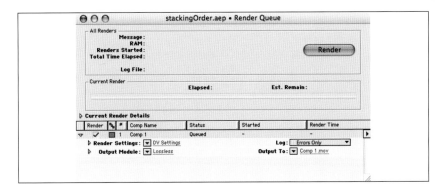
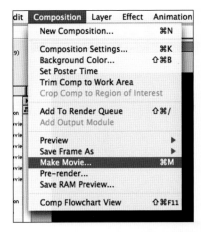

The Render Queue window enables you to set the render settings for your composition and to determine where After Effects will output the rendered file once it's finished. (For a full description of all the Render Queue's options, and there are many, see "Using the Render Queue window" in the After Effects Help files.)

> ——— NOTE ———
>
> *When you render for the first time, the Output Movie To window opens, prompting you to choose the location where your rendered file will be saved.*

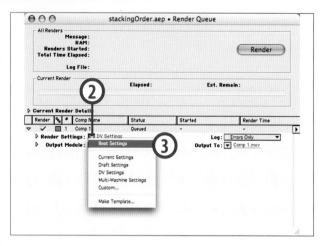

2. Click the triangle to the right of Render Settings. The Render Settings pop-up menu opens. This menu enables you to select a rendering preset.

3. For Quality, select Best Settings. This selects a render preset optimized for creating the highest quality video files. Then click OK.

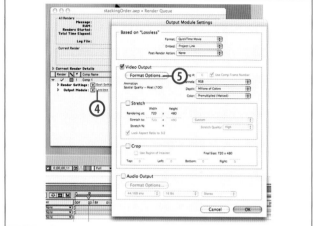

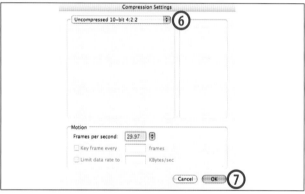

4. Click the word Lossless to the right of Output Module. The Output Module pop-up window opens.

5. Click Format Options.... The Output Module window stays open, and the Compression Settings window opens on top of it. You can now choose the exact type of compression After Effects applies to your clip.

6. Select Uncompressed 10-bit 4:2:2 from the pull-down menu at the top of the window. This is the highest quality compression setting available in After Effects.

7. Click OK. After Effects closes the Compression Settings window. The Output Module Settings window remains open.

8. If your composition contains audio, click the Audio Output option. Clicking this checkbox exports audio along with the video in your render file. If your composition contains audio and you don't check this box, After Effects will export the video only—leaving your work silent. (In the example, I left the box unchecked because the exercises in

this chapter don't use any audio. For more on using audio in your work, see Chapters 17 and 18.)

9. Click OK to close the Output Module Settings window.

10. Click the filename to the right of Output To.

After Effects automatically names the file with the name of the composition being rendered, and the selected file type. In this case, the composition is composition1, and I'm rendering it as a QuickTime movie, so After Effects has named the file *composition1.mov.*

When you release the mouse, the Output Movie To window opens. You can use this window to determine the output location of your finished render file.

11. Navigate to a location on your computer where you would like After Effects to place your file once it's rendered. In the example, I'm outputting the file to a folder on my hard drive called *rendered files.*

If you'd like to change the name of your file, you can rename it in the Save As field (just don't change the filename extension *.mov*—it identifies the file as a QuickTime movie, and you may encounter difficulties if you remove it).

— N O T E —

Rendering a composition in After Effects creates a new video file. You can store this video file in the same location as your project file and media files, or in a completely separate location. If you have more than one hard drive installed on your system, and one is starting to fill up, directing After Effects to output your rendered video file to another location can help you avoid the problem of an overfull disk.

12. Click Save. After Effects selects the location you've chosen as the destination for your rendered file and closes the window.

13. Once you've made your selections for Render Settings and Output Module, and set the location of your final output file, click Render.

After Effects begins to render your file, and displays both how much time has elapsed since you started rendering and an estimate of how much time remains.

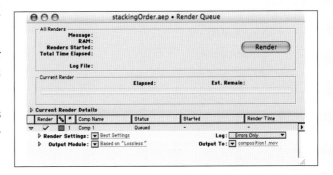

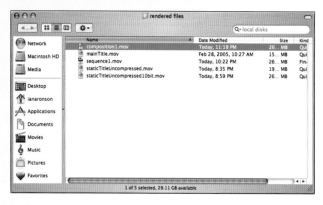

14. When After Effects has completed rendering, navigate to the folder on your computer that you selected as the output location in step 10. (In the example, I selected a folder named *rendered files*.) The folder now contains a rendered QuickTime movie of your composition.

You can bring this movie to a high-end postproduction facility to create a broadcast-quality master, or even a film print. You can also work with a compression program to encode your work for presentation on a DVD. (For more on output strategies, see Appendix A.)

Creating Images in Photoshop for Use in Digital Video

Adobe Photoshop is a tremendously valuable tool for independent filmmakers working in DV, because it allows you to add professional-quality graphics to your projects. However, because computers and televisions display images differently, ensuring that your graphics look the way you want takes some finesse and a little planning.

As described in the last chapter, an NTSC monitor displays a more limited range of color than a computer screen. Bright whites and dark blacks may look great on a computer screen, but they just won't display well on TV. Very dark objects lose detail in video if the color value exceeds the broadcast-safe range of colors a television can safely reproduce. (Broadcast-safe color is explained in detail in Chapter 16.) If dark objects appear against dark backgrounds, their edges can blend together so that you can't make out where one shape ends and the other begins. Bright whites can be equally challenging. The edges of bright white text blur if their color strays too far outside the broadcast-safe range, and can even bleed into the audio signal of a project and create a distracting hum. (You may even have noticed this while watching poorly produced commercial spots on late night TV.)

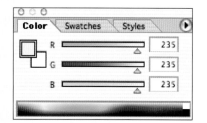

Figure 11-13. A broadcast-safe white.

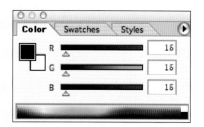

Figure 11-14. A broadcast-safe black.

One straightforward technique to help you create broadcast-safe images in Photoshop is to manage the mathematical values of the colors you choose. Each color channel in an RGB image has a maximum value of 255 and a minimum value of 0. As mentioned earlier in this chapter, setting all three sliders to 255 creates white; setting all three to 0 creates black. Both of these color combinations fall outside the range an NTSC monitor can reliably reproduce. To create broadcast-safe colors, it helps to stay away from either edge of the spectrum. Instead, to produce a broadcast-safe white, you can set the R, G, and B channels to 235 (Figure 11-13); to produce a broadcast-safe black, set each channel to 16 (Figure 11-14). You can set the value of a color you use in Photoshop by dragging the RGB sliders or by entering the numerical values directly in the text boxes on the right side of the palette. A broadcast-safe white may look slightly gray on a computer screen, but will display as actual white on an NTSC monitor.

Video also has trouble displaying bright reds, yellows, and greens, because they can bleed into surrounding areas of color (see Chapter 16). As a result, in addition to watching your white and black levels, it's a good idea to use the NTSC colors filter in Photoshop before you import any images into your video.

Applying a broadcast-safe color filter in Photoshop

To apply a broadcast-safe filter:

1. Compose and edit your Photoshop image so it looks the way you would like it to appear in your video.

2. Select Filter→Video→NTSC Colors. Photoshop adjusts the colors in your image to broad-cast-safe NTSC values.

3. Save your work.

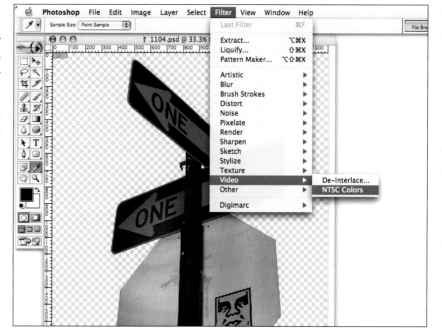

Compensating for non-square pixels

In addition to color considerations, it's helpful to think about pixel aspect ratios when you're creating graphics. (For an introduction to aspect ratio, see Chapter 2.) The pixel dimensions of a frame of NTSC mini DV video are 720×480. However, computers use square pixels to generate images, while video editing programs use non-square pixels that are taller then they are wide.

> --- N O T E ---
>
> *The difference between square and non-square pixels is described in detail in Chapter 2. The same chapter also explains the difference between video monitors, which use lines of resolution to display images, and computer monitors, which use pixels.*

After Effects automatically resizes a 720×480 image to compensate for square pixel images created in Photoshop and imported into a DV project. (Earlier I wrote that Photoshop and After Effects are engineered to work together; this is a perfect example.)

Resizing for Final Cut Pro

Unlike After Effects, Final Cut Pro does not automatically resize an image at its correct proportions.

As a result, if you create a 720×480 pixel image in Photoshop (Figure 11-15), it will distort when you import it into Final Cut Pro (Figure 11-16) and display as an oval.

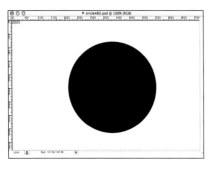

Figure 11-15. A 720x480 Photoshop image.

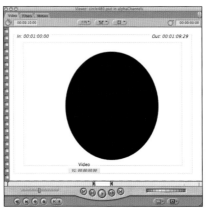

Figure 11-16. Photoshop images may distort in Final Cut Pro, because Final Cut Pro does not resize images automatically.

The solution is to create Photoshop graphics at slightly taller dimensions than the NTSC screen. This compensates for the difference in pixel aspect ratios. By creating graphics at 720×534 pixels, you can create images that maintain their proportions when you import them into a video project in Final Cut Pro or DVD Studio Pro. Photoshop even offers a preset you can apply when creating a new document.

SIDEBAR

Video Formats and Frame Size

There are a number of different NTSC video formats, and they don't all use the same size frame, which can be confusing. The most popular prosumer format is mini DV, which is used by cameras such as the Sony VX2100 and the Canon XL-2, as well as many consumer-level cameras. Another prosumer format, called DVCAM, is used by higher-end prosumer cameras such as the Sony PD-170. The mini DV and DVCAM formats share a screen size of 720x480 pixels.

If you're working with prosumer equipment, this is the frame size you'll be working with most often, if not exclusively. You can compensate for square pixels in your mini DV or DVCAM still images by working at 720x534.

However, 720x480 is not the only possible screen size. Professional video formats, such as Betacam SP and DigitalBetacam, use a screen size of 720x486 pixels. If you're working with 720x486 pixel video, you can compensate for square pixels in your still images by working at 720x540.

Because of the difference in frame size, Photoshop offers different sized presets. The 720x540 Std. NTSC 601 Photoshop preset works with professional formats like Betacam SP and DigitalBetacam. The 720x534 Std. NTSC DV/DVD preset works with mini DV or DVCAM formats. (The preset contains the word DVD, because this screen size is also appropriate for still images used in prosumer DVD creation programs such as Apple's DVD Studio Pro.)

Creating a Photoshop image for use in Final Cut Pro

To create a Photoshop image for use in Final Cut Pro, do the following:

1. In Photoshop, select File→New. The New window opens.

2. Enter 720 in the Width field, 534 in the Height field, and 72 in Resolution. Be sure to choose RGB color from the Mode pull-down menu. Alternatively, you can select 720×534 Std. NTSC DV/DVD from the Preset Sizes pull-down menu.

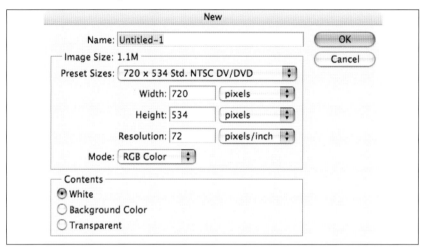

3. Click OK. Photoshop opens a new document at the dimensions you specified.

4. Create your image. A circle makes a good test, because you can easily see if it distorts after importing it.

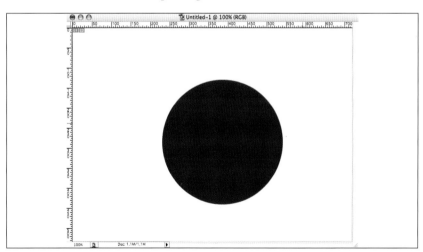

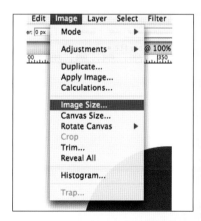

5. To compensate for the difference in pixel aspect ratio described earlier in this chapter, select Image→Image Size.... The Image Size window opens.

6. Enter "480" in the Height field. Be sure the Constrain Proportions box is unchecked.

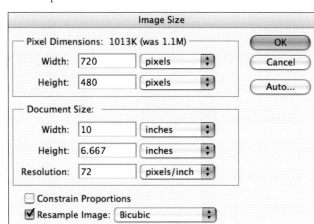

7. Click OK. Photoshop resizes the image.

Because Constrain Proportions was unchecked, the circle distorts into an oval in the resized version. When you import the image into Final Cut Pro, it will appear as a circle due to Final Cut Pro's use of non-square pixels.

8. Save your Photoshop image.

9. Import the image into Final Cut Pro. The image once again appears as a circle.

10. Save your work.

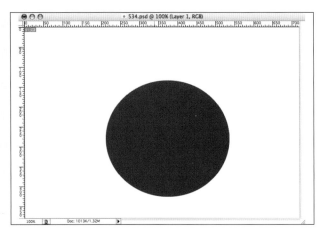

At this point, you've learned how to create professional-quality image sequences, how to render them, and how to export and composite a finished sequence. All composite effects build on the foundation material addressed in this chapter: RGB, alpha channels, stacking order, the ability to import graphics created in other software applications, and the ability to export a finished product. The chapters that follow will walk you through the process of creating complex composite effects and integrating them into your work.

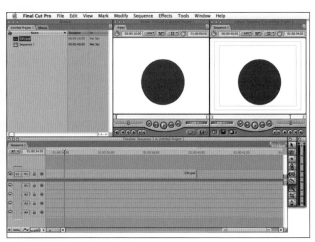

Compositing Techniques to Make Your Project Look Like It Cost More Than It Really Did

12

As explained in Chapter 11, you can combine and arrange multiple layers of video and still images to exercise complete control over all the visual material in your project.

This chapter and the three chapters that follow (Chapters 13, 14, *and* 15) walk you through the creation of an effects-intensive short film I made for this book, titled *Big Luca*. This *Godzilla* inspired sci-fi action piece (also described in Chapter 5) uses all the techniques detailed in this book. The visual climax of the film is a series of composite sequences in which a giant two-year-old towers over the Lower East Side. These shots not only serve as a punch-line to the film, but recreating them as you read this book provides you with an opportunity to practice your technique.

This chapter shows you, step-by-step, how to:

- Use a video clip as the background layer in a composite sequence.

- Remove the green screen background of a video layer, enabling the clip underneath to show through. (For details on how to set up a green screen shoot, see Chapter 5.)

- Use the luminance values in a video clip to create areas of transparency. (For more on luminance, see "Using a luma key effect to create transparency" in this chapter, as well as Chapter 16.)

- Create a matte to display specific, carefully selected areas of a video clip while the rest of the layer remains transparent.

NOTE

To recreate the effects sequences illustrated in this chapter:

- *Navigate to the ch12 folder inside the Exercises folder on the StartToFinish DVD that came with this book. Drag the entire folder to your hard drive (remember, don't try to work with video files directly from the DVD; see Chapter 10).*

- *Follow the step-by-step instructions in each section.*

The ch12 folder on the DVD also contains finished Final Cut Pro and After Effects project files that you can use as a reference.

Creating the Houston Street Composites

The composite sequences *Big Luca* is built around were shot in two parts. In one shoot, Director of Photography Daniel Baer and I recorded a series of exterior shots on East Houston Street in Manhattan to provide buildings and traffic for Luca to tower over (Figure 12-1). In the other shoot, we set up a green chroma key background in the apartment that Luca shares with his parents and two younger brothers (Figure 12-2). These interior shots provided us with images of Luca to composite into the exterior clips. Combining the shots into finished composite sequences (Figure 12-3) required the full range of key effects and masking techniques detailed in this chapter.

Figure 12-1. An exterior shot of Houston Street.

Figure 12-2. An interior green screen shot.

Figure 12-3. A screen capture of the finished composite.

The final composite sequence uses multiple layers of video to make Luca appear as if he's actually in the environment. He's composited on top of one layer of video (so the sky appears behind him) and beneath another layer (so the buildings appear in front of him). Using this many layers of video together creates a more involved postproduction process, but ultimately leads to a much stronger sequence. The first step is creating a new sequence in Final Cut Pro, or a new composition in After Effects, and then adding a background layer of video.

Building a sequence, and creating a background

To build the sequence in Final Cut Pro, do the following:

1. Select File→New→Sequence. A new sequence, named Sequence 1, appears in the Browser window. Name and save the project.

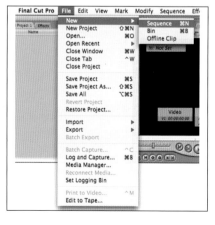

NOTE

It's a good idea to start saving right away, and to save periodically as you go along. That way, if the computer crashes, you won't lose your work.

2. Click the name of the file once to select it. The sequence name appears highlighted in the Browser window.

3. Click the name again. The color changes and a cursor appears in the name. You can now give the sequence a new name by typing on your computer's keyboard. (Giving each sequence a descriptive name can help you keep track later as your project grows and becomes more complex.)

4. Name the sequence houston1, and press return.

5. Double-click the houston1 sequence. Final Cut Pro opens the sequence in the Timeline.

6. Select File→Import→Files. Navigate to the *ch12* folder on your hard drive.

7. Click the file at the top of the list, *houston1.mov*, then shift-click the file at the bottom of the list, *luca3.mov*. This selects all the files in the folder.

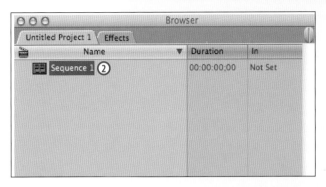

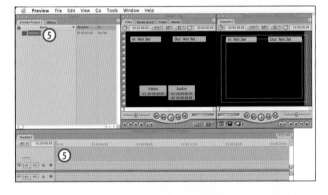

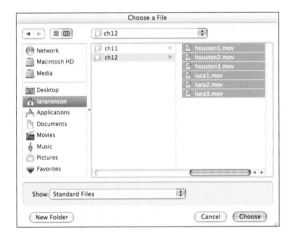

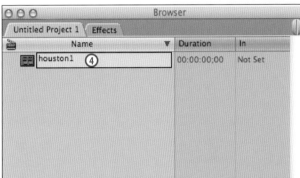

8. Click Choose. This imports the files (they now appear in the Browser) and closes the window.

9. Drag the clip *houston1.mov* into track V1 in the Timeline. When you release the mouse, Final Cut Pro adds the clip to the Timeline and displays a frame of the clip in the Canvas.

10. Save your work.

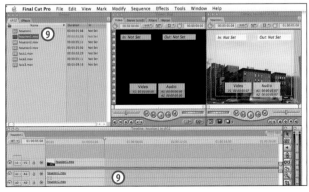

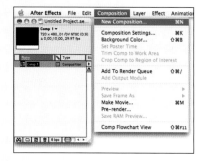

To build the sequence in After Effects, do the following:

1. Select Composition→New Composition. The Composition Settings window opens.

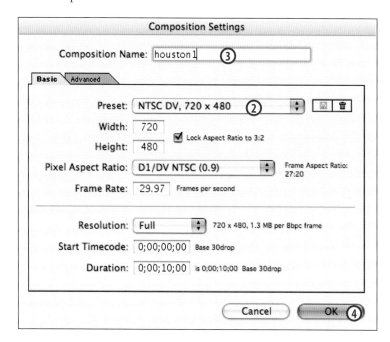

2. Select NTSC DV, 720×480 from the Preset pull-down menu, or if you're working in another video format, pick the appropriate pull-down menu option (see Chapter 11 for more detail).

3. Click in the Composition Name field, and type houston1 on your keyboard. After Effects names the composition houston1. (If you don't enter a name now, After Effects will automatically name the composition Comp1. As with a sequence in Final Cut Pro, giving an After Effects composition a more descriptive name will help you keep track as the project develops and you start to create additional compositions.)

4. Click OK. After Effects creates a new composition and closes the window.

5. Select File→Import→File.... The Import File window opens. (You can also open the Import File window by double-clicking an empty area inside the Project window.)

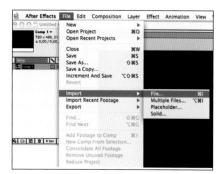

6. Select the following files:

   - *houston1.mov*

   - *houston2.mov*

   - *houston3.mov*

   - *luca1.mov*

   - *luca2.mov*

   - *luca3.mov*

   (The easiest way to select multiple files together is to click *houston1.mov*, and then shift-click *luca3.mov*—After Effects will automatically select contiguous files.)

7. Click Open. After Effects imports the files and closes the Import File window.

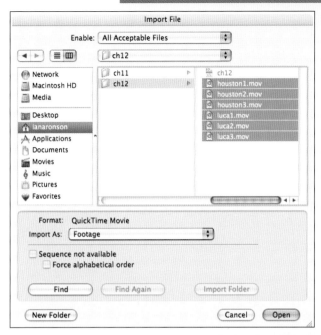

NOTE

After Effects offers another option, labeled Import Multiple Files, which is helpful when you want to import files from multiple locations. For example, if you had one file on your computer's desktop, another in a folder on your computer's hard drive, and a third on an external FireWire disk, selecting Import→Multiple Files would be the way to go. If you choose this option and then import a file, After Effects will automatically open a new Import File window. This new window enables you to then navigate to another location and import additional files. However, if you're importing multiple files from the same folder, selecting File→Import→Multiple Files is unnecessary. For the example in this chapter, shift-clicking is the simplest route you can take.

8. Select *houston1.mov* in the Project window, and drag it into the Timeline. After Effects places the file in the Timeline and displays a frame of the clip in the Composition window.

9. Save your work.

Scaling the background clip

As described in Chapter 5, an object with cleaner edges is easier to composite into another clip. Objects with fine detail at their edges (Figure 12-4), for example shaggy wardrobe materials, bushy hair on an actor, or the leafless tree branches on the right and left edges of the *houston1.mov* clip, are much harder to work with in a key effect. Scaling the clip to make it slightly larger enables us to crop out the edges of the shot which contain the problematic tree branches (Figure 12-5).

Although there's no key effect applied to this layer of the composite, later in the chapter, we'll use an effect to create transparency in another instance of the same clip placed on a higher level video track. Scaling both the top and bottom layer clips to the same size makes them easier to composite together.

Final Cut Pro and After Effects both enable an editor to scale a clip by clicking in the Canvas (Figure 12-6) or Composition window (Figure 12-7) and dragging a corner of the clip. Dragging a corner toward the center of the window scales the clip down, reducing its size. Dragging a corner toward the outside of the window scales the image up to a larger size.

Figure 12-4. The tree branches in this shot make the clip more difficult to composite.

Figure 12-5. The same shot, scaled to make the branches much less prominent.

Figure 12-6. A selected clip in the Final Cut Pro Canvas, with scale handles in each corner.

Figure 12-7. A selected clip in the After Effects Composition window, with scale with handles in each corner.

This method is imprecise, and in After Effects, it can distort the image if you change the aspect ratio (see Chapter 2).

Scaling an image numerically enables you precisely control its size. In this case, numeric scaling is especially helpful because the composite effect in this chapter places the same clip on more than one layer, and they both need to appear at the same size for the effect to work. (Plus, it adds another skill to your repertoire and helps you develop your technical chops as an editor.)

SIDEBAR

Action Safe, Title Safe, and Underscan

Unlike a computer monitor, a television screen doesn't display a full frame of video. Televisions automatically crop video images to show a smaller area of the frame. (The amount of cropping varies from monitor to monitor.) This means that although objects at the edge of the frame may be visible in the Final Cut Pro Canvas or After Effects Composition window, they might not be visible to an audience watching your work on a TV screen.

Professional video equipment is designed to show an entire frame of video so you can spot any possible problems with your material (the process is called underscanning). To help estimate what an audience will see on a video monitor, Final Cut and After Effects both use action-safe and title-safe boundary lines.

- The action-safe area is 10 percent smaller than a frame of video.

- The title-safe area is 20 percent smaller than a frame of video.

monitor to your editing system, see "The importance of a well-calibrated NTSC monitor" in Chapter 10.)

Because different television monitors display slightly more or less of the video frame, both the title-safe and action-safe areas are approximations. Images that appear inside the action-safe area will more than likely appear on a television screen. Images inside the title-safe area will definitely appear. Because titles and visual effects lose their value if audiences can't see them, it's important that their edges don't get cut off. Framing your composites with an eye toward the action-safe and title-safe areas is a good way to make sure your viewers will be able to see the composites you've worked so hard to create.

Again, the title-safe and action-safe areas are approximations. To really see what your work looks like on a television screen, it helps to view your work on a dedicated NTSC video monitor. (For more on connecting a video

To display the action-safe and title-safe boundaries in After Effects, click the Title-Action Safe button at the bottom of the Composition window (it's the third button from the left in the image above).

To display the action-safe and title-safe boundaries in Final Cut Pro:

1. Select View→Show Overlays.

2. Select View→Show Title Safe.

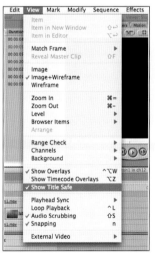

Numerically scaling an image in Final Cut Pro

To scale an image numerically using Final Cut Pro, do the following:

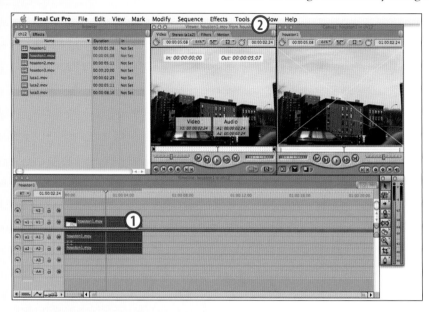

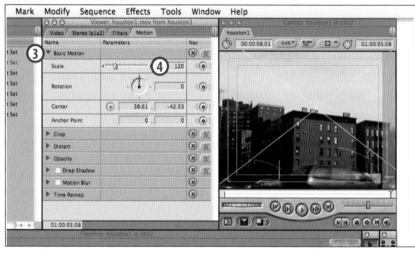

1. Double click the houston1 clip in the Timeline. Final Cut opens the clip in the Viewer window.

2. Click the Motion tab in the Viewer. Final Cut Pro displays a series of controls for the clip.

3. Click the triangle to the left of Basic Motion to access the Basic Motion controls. The Basic Motion controls become visible.

4. Enter 120 in the Scale field and press the return key. Final Cut scales the clip to a larger size, 20 percent larger than normal, and displays the resized clip in the Canvas. The clip is now larger than a frame of video, and anything that does not appear in the Canvas will be cropped out of the frame.

5. Click in the center of the Canvas, and drag the clip up and to the right (so you can frame out as much of the trees as possible, while still retaining the traffic motion in the bottom of the frame).

6. When you're satisfied with the position of the enlarged clip, release the mouse. The Center fields in the Viewer display the clip's new coordinates. (The default center position of a clip is 0,0: the center of the frame.) The number in the first Center field is the clip's X, or horizontal, coordinate. The number in the second field is the Y, or vertical, coordinate.

7. Save your work.

Numerically scaling an image in After Effects

To scale an image numerically in After Effects, do the following:

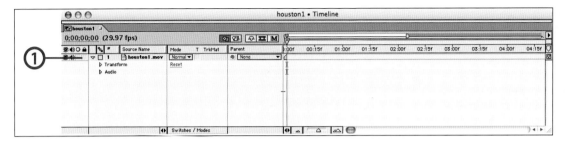

1. Click the small triangle (sometimes referred to as a twirl-down icon) on the first layer in the Timeline. After Effects opens the layer's controls.

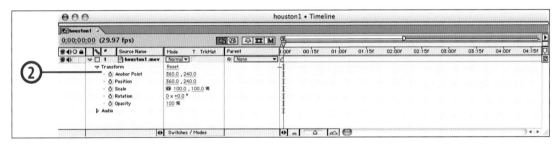

2. Click the small triangle next to Transform. After Effects opens the Transform controls for the clip on layer1.

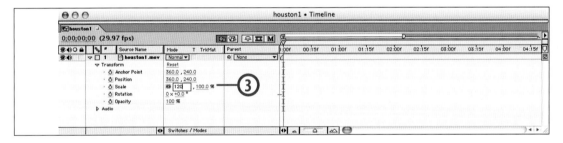

3. On the Scale line, click the first number (100.0). The field changes color and a cursor appears. This is the horizontal scale control for your clip.

4. Type 120 on your keyboard and press return. By default After Effects constrains the proportions of your clip, so when you change the horizontal scale value the vertical scale automatically changes with it.

5. Click in the Composition window and drag the image up and to the right. In the last step you made the image 20 percent larger. Because the image is now larger than a frame of video, anything that doesn't appear in the Composition window will be cropped out of the frame.

6. When you're satisfied with the position of the enlarged clip, release the mouse. The Position fields in the Timeline display the clip's new coordinates. The number in the first Position field is the clip's X, or horizontal, coordinate. The number in the second field is the Y, or vertical, coordinate.

7. Save your work.

Adding a new clip and applying a chroma key effect

The next step in building this composite sequence is to add a new clip, and then remove new clip's background so that Luca appears in the exterior shot of Houston Street. To make this effect possible, we shot Luca against a green background, which was designed specifically for use in chroma key effects. (For more information on chroma key, see Chapter 5.) As a result, you can now cleanly remove the green background using your digital editing system.

The first step is adding the new clip to your existing sequence.

Adding a new video track and editing a new clip into a Final Cut Pro sequence

Although you can easily add a new video clip to a Final Cut Pro sequence by selecting the clip in the Browser and dragging it into the Timeline (as demonstrated in Chapter 11), there is another way that provides you with greater control. This exercise shows you how edit a new clip into the Timeline at the exact location you want, using the Playhead, Source, and Destination controls.

1. Click in the Timeline and select Sequence→Insert Tracks.... The Insert Tracks window opens

2. Type 1 in the Video Tracks field in the Insert Tracks dialog box. Leave the After Last Track radio button selected (this will add a new video track to the Timeline, above the video track containing the exterior clip you added earlier). Click OK when you're done.

3. Double click the clip named *luca1.mov* in the Browser. The clip opens in the Viewer. This is the clip you will edit into the Timeline.

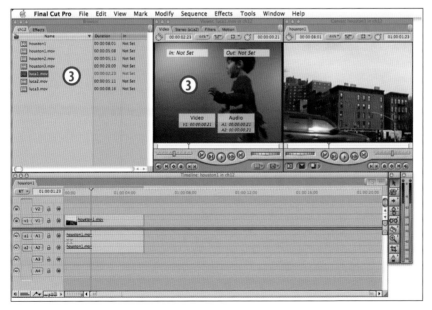

4. Drag the Source video tab (labeled v1) up to the Destination video tab (labeled V2), which is next to the empty video track in the Timeline. Your source clip is the footage of Luca displayed in the Viewer. Your destination is the empty video track.

5. Drag the audio Source tabs (a1 and a2) to align with the audio Destination tabs for the empty audio tracks in the Timeline (A3 and A4).

6. Drag the Playhead in the Canvas until the black SUV appears about halfway through the frame. When you drag the Playhead in the Canvas, the Playhead in the Timeline moves along with it. This enables you determine exactly where the first frame of your new clip will appear in the Timeline.

7. Click the Overwrite edit button in the Canvas. Final Cut adds the new video clip to track V2 in the Timeline and adds the associated audio clips to tracks A3 and A4, as you specified. The audio and video of the exterior shot on tracks V1, A1, and A2 remains undisturbed.

── N O T E ──

Because the Playhead was positioned almost two seconds after the start of the clip when you performed your edit, Luca does not appear at the very beginning of the shot. Introducing Luca and the chroma key effect part way through the shot gives it greater impact and an added degree of surprise. The audience will get used to watching an ordinary street scene, and then once they've gotten comfortable with the shot, a giant two-year-old will suddenly appear.

In filmmaking, just like in telling a good joke or starting a new romance, timing is everything.

Adding the chroma key effect in Final Cut Pro

Now that you've added Luca to the Timeline, you're ready to remove the green screen background and composite him into the exterior video clip. In this exercise, you identify the precise hues, or mathematical color values, of the green screen background and remove them from the shot (for more information on hue, see Chapter 16). As you remove areas of color, the layer of video underneath begins to show through.

1. Select the *luca1.mov* clip in the Timeline. The clip changes color and becomes the active clip.

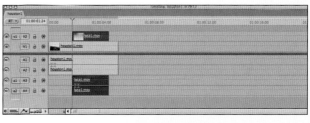

2. Select Effects→Video Filters→Key→Chroma Keyer. Final Cut adds a chroma key effect to the clip, but you won't see any changes until you set the parameters in the following steps.

3. Double-click the clip in the Timeline. The Chroma Keyer tab appears at the top of the Viewer window. Click the Chroma Keyer tab to access the controls. Using these controls, you can key out, or remove, the green background.

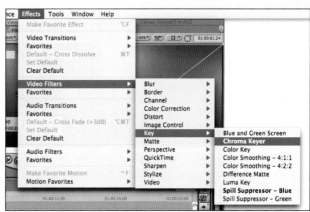

4. Click the Eye-dropper tool and use it to select an area of color in the Canvas. When you click with the Eye-dropper tool Final Cut Pro identifies the color in the part of the frame that you click on and eliminates that specific color from the shot. Video from the layer beneath then begins to show through.

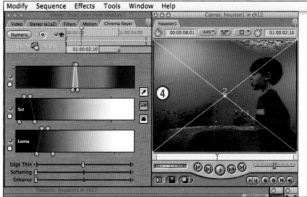

NOTE

In an ideal scenario, you would be working with a perfectly lit solid colored background, and you would only need to click once to remove the entire green screen backdrop. This clip, as in most chroma key clips you'll encounter as an editor, contains multiple shades of color and requires more than a single click to remove them. As a result, it provides a great opportunity for you to practice.

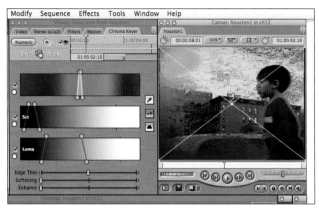

5. Click the Eye-dropper tool again, hold down the Shift key, and click a new area of the green screen background. Final Cut removes additional areas of color from the shot.

NOTE

When you're working with chroma key effects in Final Cut Pro, as in this example, you may notice a harsh or pixilated edge appears around the area of a shot that has not been keyed out. You can fix this by using the Softening control in the Chroma Keyer effect. The Softening control blurs the edges of the key effect and helps blend the top layer clip into the clip underneath.

6. Repeat step 5 until you've removed the entire background.

7. Drag the Softening slider slightly to the right. Final Cut Pro softens the edge of the key around Luca. Once the edges no longer look harsh or jagged, you can stop dragging. (If you drag the slider too far to the right, Luca begins to look like a ghost without solid edges.)

8. Save your work.

Adding a new clip into an After Effects Composition

To add a new clip in After Effects, do the following:

1. Scroll through the clip using the Current Time Indicator until the black SUV becomes visible in the frame (see the note about timing in "Adding a new video track and editing a new clip into a Final Cut Pro sequence"). The Current Time Indicator in After Effects functions much like the Playhead in Final Cut Pro or Macromedia Flash: when you edit a new clip into the composition in the next sequence, the first frame will appear at the Current Time Indicator's present location.

2. Select the clip you'd like to add to the composition and press Command+/ (Mac) or Ctrl+/ (Windows). You can also drag the clip you'd like to add directly into the Timeline (as you did earlier in this chapter) or into the Composition window.

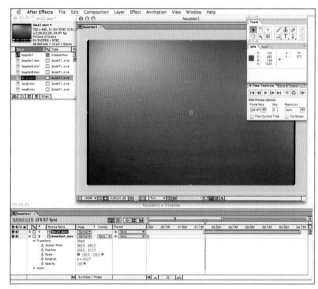

The new clip appears on a layer above the Houston Street clip.

Adding the chroma key effect in After Effects

To create a chroma key effect using After Effects, do the following:

1. Click the *luca1.mov* clip in the Timeline to make sure it's selected.

2. Select Effect→Keying→Color Range. The Effect Controls window opens, displaying the parameters of your key effect. The clip itself remains unchanged until you set the parameters of the effect.

3. Click the Eye-dropper tool in the Effects Controls window, and then click an area of the clip in the Composition window. After Effects identifies the color you click on, and then removes areas of that color from the shot—the layer of video underneath begins to become visible.

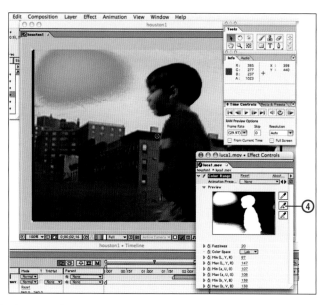

4. Click the Eye-dropper tool with a + next to it, and click another area in the Composition window. After Effects removes additional areas of color from the clip, and an increasing area of the base video layer becomes visible.

5. Continue to key out areas of color until the green background becomes fully transparent.

The Preview in the Effect Controls window displays a black and white image that changes as you select areas to key out. This black and white image is a matte, and determines what parts of your clip become invisible. The black areas are parts of the *luca1.mov* clip that are set to transparent, and the white areas are parts of the image that remain opaque.

Depending on the footage you're working with you may find that the foreground image you're working with—in this case, Luca—appears to be surrounded by a noticeably jagged edge. If you're using the professional version of After Effects, you can compensate for this by adjusting your matte with the Matte Choker Effect.

6. Select Effect→Matte Tools→Matte Choker. The parameters of your Matte Choker effect appear in the Effect Controls window.

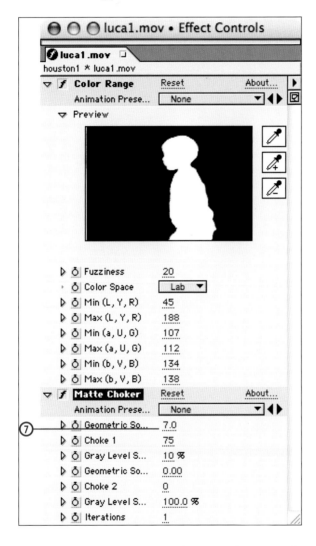

7. Click the field next to Geometric Softness, and enter 7. This softens the edge of the matte you created while making your key effect and helps your foreground clip, Luca, blend more easily into the exterior shot.

8. Save your work.

Using a luma key effect to create transparency

As described earlier in this chapter, there is more than one type of key effect available to you as an editor. Similar to a chroma key effect, which identifies a particular color value and removes it form a shot, a luminance or luma key effect identifies a particular brightness value and removes that value from a shot. Luma keys are very useful in removing large, irregularly shaped areas that are significantly brighter or darker than other image areas in the same clip.

At this point, Luca appears to be playing in the traffic on Houston Street. The chroma key effect has cleanly separated Luca from the green-screen background, but he doesn't look especially big at this stage of the composite. Objects that are closer to the camera generally appear larger than objects that are farther away, so it makes sense that a boy in the foreground would appear larger than the cars behind him. (In real life, Luca is very big for his age, which is part of what inspired this film, but he's not Godzilla size.) To complete the effect, we need to add a layer of buildings in front of Luca, and create an area of transparency so Luca can look like he's towering over the street scene.

In this exercise, you'll add a new video clip to create a layer of buildings in the foreground, and use a luma key effect to make the sky transparent. Because the sky is significantly brighter than other parts of the shot (even on a cloudy day) the luma key effect works nicely.

Creating a luma key effect in Final Cut Pro

To create the luma key effect, do the following:

1. Add another instance of the *houston1.mov* clip to a new track in the Timeline (you can either drag the clip from the Browser into the Timeline or use the Overwrite edit key).

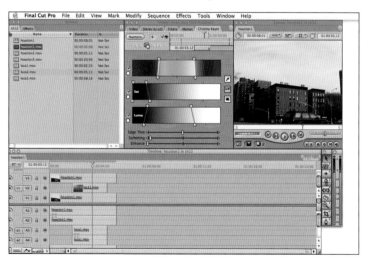

2. Double-click the *houston1.mov* clip you just added to track V3 in the Timeline. The clip opens in the Viewer.

3. Click the Motion tab to display the clip's motion controls.

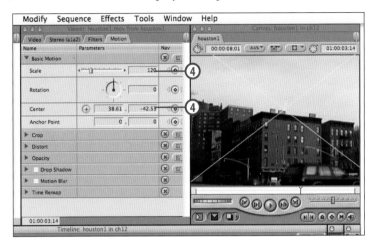

4. Set the Scale and Center controls to match the clip on track V1:

   • Enter 120 in the scale field.

   • Enter 38.61 in the first Center field and −42.53 in the second.

5. Select Effects→Video Filters→Key→Luma Key. Final Cut Pro applies a luma key effect to the clip.

6. Click the clip's Filters tab in the Viewer. Final Cut displays the controls for the Luma Key effect. At first, when the effect's default settings are applied, Luca looks like a translucent ghost. The entire top layer street scene clip on track V3 is now translucent, and Luca is now dimly visible behind it.

7. Drag the Tolerance slider all the way to the left. Setting the Tolerance to 0 makes Luca clearly visible behind the buildings. The Threshold slider, which you'll adjust in the next step, defines the luminance value that Final Cut Pro removes from the shot. The Tolerance slider defines how similar luminance values will be handled: a higher Tolerance setting removes a wider range of luminance values, while a lower range removes a much narrower range, leaving the rest of the image intact.

8. Drag the Threshold slider to the right (in the illustration, the Threshold is set to 38). Areas of a shot containing bright reflections, such as windows or the shiny roof of a car, often share the same bright luminance values as the sky. If you leave the Threshold at 0, some of the windows in the buildings may get keyed out and become transparent, allowing Luca to show through behind them.

9. Save your work.

Creating a luma key effect in After Effects

To create the luma key effect in After Effects, do the following:

1. Add another instance of the *houston1.mov* clip to the Timeline.

2. Scale the clip to match the instance on the bottom layer:

   • Enter 120 in the scale field and press return.

   • Enter 402.0 and 217.0 in the position fields.

3. Select Effect→Keying→Luma Key. The parameters of your luma key effect appear in the Effect Controls window. The trick to working with luma key is to simultaneously adjust all four of its settings, which are briefly described as follows (Figure 12-8):

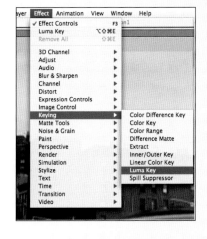

Threshold

    This setting defines the luminance value that After Effects makes transparent.

Tolerance

    This slider defines how similar luminance values will be handled: a higher Tolerance setting removes a wider range of luminance values, a lower value removes a much narrower range leaving the rest of the image intact.

Edge Thin

    This sets the width of the border, depending on your settings, for the matte created by the luma key effect.

Edge Feather

    This softens, or blurs, the edge of the matte. This blends the top layer video clip more easily into the layer below.

In the example, Threshold is set to 148, Tolerance to 0, Edge thin to 0, and Edge Feather to 6.7. All four of these controls work together to change your composite. As you work on other clips in the future, adjust each of the four parameters in combination to find the settings that work best for the particular composite you're attempting.

Figure 12-8. The composite sequence with a completed luma key effect.

Drawing a mask in After Effects

You now have an entirely usable composite sequence. (Give yourself a pat on the back.) You've combined three separate clips into a single composite image, and you've created and adjusted areas of transparency to create an intricately layered effect. If you'd like to polish this effect and take it to a more professional level, there's one more step. You can add a new layer of video, and then add a mask.

A mask defines which part of a frame will be transparent and which part will remain fully visible. The best part of creating a mask in After Effects is you can use the Pen tool to simply draw the mask in the exact shape you want.

You may have noticed a slight halo, or glow, around the edge of the luma key matte that you softened with the Edge Feather slider. You also may have noticed that Luca slightly shows through some of the brighter areas in the foreground Houston Street clip (for example, in Figure 12-9, you can see the red of his shirt bleeding faintly through some of the windows in the building in front of him, and through the reflection on the roof of the taxi at the bottom of the frame).

If you add a new Houston Street clip to the composition, and draw a mask along the top of the buildings, you can create a composite with a very clean edge (Figure 12-10). The buildings and traffic will show through as solid objects in the bottom half of the frame, and the sky will become completely transparent. (In some ways, this technique provides an alternative to the luma key used earlier. Once you get comfortable drawing masks, you can use them to create transparent areas of any shape, in any shot of video.)

Figure 12-9. The composite sequence before a mask has been applied. Notice the soft edge at the top of the buildings.

Figure 12-10. The composite sequence after a mask has been added, the top edges of the buildings appear much cleaner.

The following steps walk you through drawing a mask in After Effects:

1. Add a new instance of *houston1.mov* to the Timeline, then resize and reposition it to match the clips you added earlier. (To streamline your production, you can delete the clip containing the luma key effect, since you don't need both. In this case, I'm leaving both in the Timeline so you can see how they work in the example file on your DVD.)

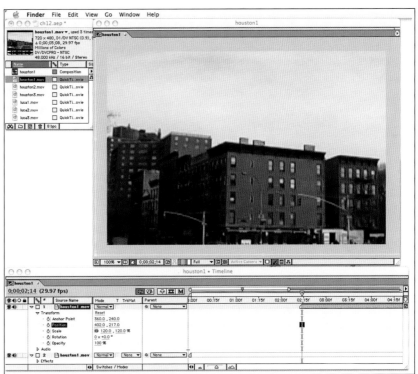

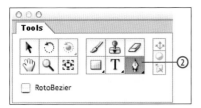

2. Click the Pen tool in the Tools palette. The cursor changes from an arrow to a pen. You create a mask in After Effects by clicking with the Pen tool. To make things easier in this exercise, make sure RotoBezier is unchecked. The RotoBezier function automatically draws Bezier curves, which you don't need in this mask and could make life slightly more difficult.

3. Click along the top edges of the buildings to create your mask. Each time you click, After Effects adds a new point to your mask and connects each point to the next.

4. Close the mask by clicking around the bottom corners of the frame, and keep adding points until you reach the first point you created. Once you close the mask by connecting

your first point and your last, After Effects makes the area of the frame outside your mask transparent. Luca now shows through above the buildings.

The soft edge of your luma key effect still shows through the frame. To see what your mask looks like without the luma key effect, you can make the layer containing the effect invisible.

5. Click the visibility icon for layer 2 in the Timeline window (it looks like an eye). Layer 2, which contains the luma key effect, becomes invisible.

6. Save your work.

Your composite effect now appears highly refined, with clean edges and several completed effects that work very nicely in combination. In this chapter, you used various scale, key, and matte effects to create a very professional sequence. (This was no small task; the exercises you completed here use some highly sophisticated techniques.)

The DVD folder for this chapter contains two additional shots of Luca in front of a green screen background (*luca2.mov* and *luca3.mov*), along with two additional street scene shots (*houston1.mov* and *houston2.mov*). If you'd like to practice these techniques further, try creating two additional composite sequences using the remaining clips. All the shots will composite together nicely using the same techniques you explored earlier in this chapter.

Chapter 13 shows you various techniques to bring still images into your project and animate them. As mentioned at the start of this chapter, you have full control over all the visual material in your project. Let's put it to use.

Artistically Using Still Images

Any film, video, or animation project is essentially just a series of stills. Each frame is a discrete still image, slightly different than the last. When quickly displayed one after the other, these stills appear to form an image that moves (as mentioned in Chapter 2, this is called persistence of vision).

Understanding the way still images function on screen gives you both tremendous control and creative freedom. Some of the first artists to take advantage of this were cartoonists. Pioneers of the animated form, such as Chuck Jones of Bugs Bunny fame, created still images on clear cells of acetate and then layered them one on top of another to create animations that functioned more or less like frames of film or video. In fact, many of Jones's techniques, including keyframes (which provide the foundation of all motion graphic programs in use today) and in between frames (which provide the basis for tweening, or intelligently filling in frames between keyframes, in animation programs like Macromedia Flash) are directly relevant to composite techniques in DV filmmaking. Especially when you're working with still images.

NOTE

There's a great documentary, Chuck Jones: Extremes and In-Betweens—A Life In Animation, *that explores Chuck Jones's work as a director at Warner Brothers, and the techniques he and other animators used to bring Daffy Duck and Elmer Fudd to life. It's fun to watch if you're into animation techniques, or even if you just want to see clips of his cartoons. If an operatic "kill the rabbit," brings back memories, this documentary is worth watching.*

Including stills in your video doesn't mean you have to create a slide show or something that looks like a PowerPoint presentation. This chapter explores ways to creatively incorporate stills into your work by animating a still image to simulate zooms and camera movements. Using your digital editing system, you can not only simplify the process of animating still images, but you can create smoother motion with precise control beyond what you could achieve with a camera and tripod.

Animating Still Photos to Simulate Camera Movements

NOTE

To recreate the effects sequences illustrated in this chapter:

- *Navigate to the ch13 folder inside the Exercises folder on the StartToFinish DVD that came with this book. Drag the entire folder to your hard drive (remember, don't try to work with video files directly from the DVD; see Chapter 10).*

- *Follow the step-by-step instructions in each section.*

The ch13 folder on the DVD also contains finished Final Cut Pro and After Effects project files that you can use as a reference.

I once worked as assistant editor on a PBS documentary directed by the extremely accomplished filmmaker, Bill Jersey. People from around the world were constantly visiting his office, where he would regale them with stories of his adventures behind the camera. He was equally talkative during production. After one very long and careful combination of pans, tilts, and zooms across the length of a very large head and shoulders portrait on the wall of a mansion, he joked that Ken Burns would have been able to make an entire film from just that shot. He was not belittling Ken Burns, but was making a reference to a style of filmmaking that Burns uses extensively in his work.

NOTE

Burns is by no means the first director to use still images in a film, it's a time-honored technique that's been in use for years. The 1957 documentary City of Gold, *produced by the National Film Board of Canada, uses still images to tell the story of the Klondike gold rush. The 1962 science fiction film* La Jetée, *directed by Chris Marker and mentioned in the sidebar "The Kitchen Conqueror and the Power of Still Images," doesn't use a single moving image—just an eerie selection of black-and-white stills and an excellent sound design. At the same time, Burns has made still images and camera movements his trademark, to the point that the controls for panning and zooming on a photo in Apple's iMovie are called "The Ken Burns Effect."*

In his epic series on the Civil War, Burns relied on still images and dramatic readings of soldiers' letters to tell his story. To keep the audience's attention and develop a visual style beyond a slide show, Burns carefully moved the camera up and down the length of the photos, zooming in on details and faces to heighten the emotion of each shot.

If you've ever tried to shoot stills, you know this is no easy task. To record a clean image, the plane of the camera lens must be identical to the plane of the photo—if the camera is slightly angled, the image will keystone and one edge will look shorter than the others. Smooth camera movements and zooms on still images are also very difficult. Shooting a still so it takes up a full frame of film or video requires a very tight zoom, which makes the slightest camera bobble especially noticeable to the audience. Even with the best equipment in the hands of a very talented cinematographer, pans and tilts can cause a camera to jerk, which is the last thing you want on screen.

To avoid these problems in the pre-digital era, filmmakers placed photos on animation stands and shot them with expensive motion control cameras that automated each movement. (In fact, controlling these moves was one of the first uses of computers in filmmaking, long before computers were used to edit.) The results were fantastically smooth but frighteningly expensive. Fortunately, like so many other aspects of film production, digital technology now makes it possible to get really good results without spending big bucks.

Digital editing software today ships with motion-control features that you can employ to move an image through a frame of video, dynamically changing its position on screen. You can even simulate complicated pans and tilts that would be exceptionally difficult—if not impossible—to create with a camera on a tripod. Using keyframes, motion paths, and the scale controls in Final Cut Pro or After Effects, you can create exactly the movement you want.

Simulating a zoom

You can easily simulate a zoom in Final Cut Pro or After Effects by changing the scale settings of a still image. In Chapter 12, we used the scale settings to resize an entire clip. In this exercise, we use keyframes in combination with scale settings, so that the image gets larger over time and fills more of the frame, to simulate a zoom.

In Final Cut Pro

Here are the steps for simulating a zoom in Final Cut Pro:

1. Select File→Import→Files. The Choose a File window opens.

2. Navigate to the *ch13* folder on your hard drive and import *zoom1068.psd*.

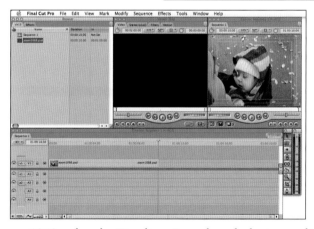

3. Add *zoom1068.psd* to the Timeline. Even though the image dimensions are significantly larger than a frame of video, Final Cut Pro scales the image to fit in the frame. (In case you haven't recognized him yet, the baby in the picture is Luca, star of *Big Luca*. This picture shows him as a pre-Godzilla–sized infant.)

Keyframes in Animation and Video

Keyframes indicate the important frames where changes take place. In the days of hand-drawn animation, lead animators like Chuck Jones would draw the keyframes of a cartoon—for example one keyframe might be Elmer Fudd pointing a gun at Daffy Duck, and the next might be Daffy with his bill pointing backward after Elmer has pulled the trigger. The less-senior animators would then draw the in-between frames, the connecting frames that create the smooth motion between keyframes. Motion graphics applications and video editing software, such as Final Cut Pro, employ similar keyframe operations. The editor sets the parameters for each keyframe, for example Scale and Center, and the computer adjusts the frames in between to create smooth, natural-looking animated effects.

Changes to a keyframe impact the keyframe itself, as well as the frames between that keyframe and the keyframes before or after it. Imagine three keyframes in a video clip, one in the first frame, another at a frame in the middle of the clip, and a third in the last frame. If the Scale setting were set to 100 percent in the first keyframe, 50 percent in the middle keyframe, and 100 percent in the last keyframe, the image would start at full size, shrink to half its size, and then grow to regain its full size. If the middle keyframe were changed to a different Scale setting, it would not impact the other two keyframes, only the frames in between. Likewise, if the first keyframe were changed, it wouldn't impact the middle or final keyframe; the change would apply only to the first keyframe, and the frames leading up to the second keyframe.

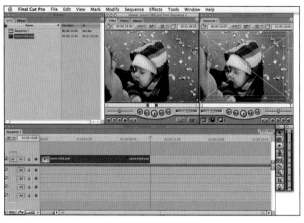

4. Double-click the image in the Timeline. The image opens in the Viewer window.

5. Click the Motion tab in the Viewer. Final Cut Pro displays the Basic Motion settings for the clip, including the Scale setting. The actual pixel dimensions of the image are 1440×1068, which is twice the size of a mini DV frame (this includes compensating for nonsquare pixels in Final Cut Pro; see Chapter 11). To fit the entire image on the screen, Final Cut Pro automatically scales the image to 50 percent of its original size.

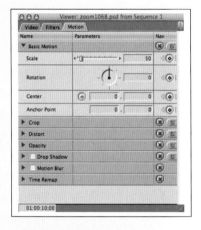

To simulate a zoom, you can set the image to go from 50 percent of original size in the first frame of the Timeline, to 100 percent in the last frame.

6. Click in the Timeline window, and then press the "home" key on your keyboard. This moves the Playhead to the first frame in the Timeline.

7. Click the Insert/Delete Keyframe button next to the Scale field in the Viewer. The diamond at the center of the button changes from white to green, and Final Cut Pro adds a keyframe to the first frame in the Timeline.

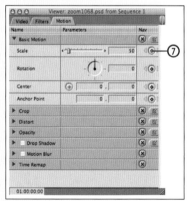

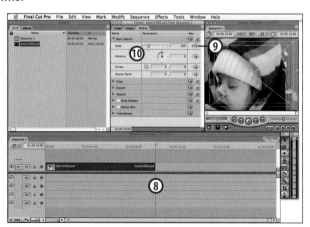

8. Click the End key on your keyboard. This moves the Playhead to the last frame in the Timeline.

9. Click the Insert/Delete Keyframe button next to the Scale field in the Viewer. Final Cut Pro adds a keyframe at the end of the Timeline.

10. Click in the Scale field, and type 100. Final Cut Pro scales the image to appear at full size. Because the image is considerably larger than a frame of video, only a tight close-up of the baby is visible.

11. Move your Playhead back to the start of the Timeline, and preview the effect you just created. Because you added keyframes before changing the Scale settings, the image now gradually changes from a reduced size of 50 percent in the first frame to 100 percent in the last frame, simulating a zoom.

12. Save your work.

In After Effects

Here are the steps for simulating a zoom in After Effects:

1. Select File→Import→File. The Import File window opens.

2. Select *zoom960.psd* and click Open. After Effects imports the file.

3. Select Composition→New Composition. The Composition Settings window opens.

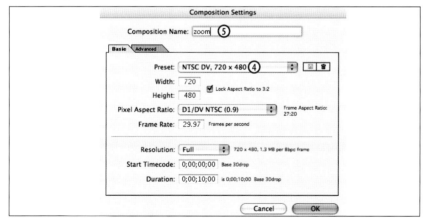

4. Select NTSC DV, 720×480 from the Preset pull-down menu (or another preset if you're working with a different video format; see Chapter 11 for more detail).

5. Click in the Composition Name field, and type Zoom. This names the composition.

6. Click OK to create a new composition and close the Composition Settings window.

> **NOTE**
>
> *Depending on your computer's processing power and the amount of RAM you have installed, you may be able to preview this effect without rendering (for more on rendering, see "Rendering performance and hardware requirements" in Chapter 10). If your computer doesn't allow you to preview the effect without renderings, you can use the arrow keys on your keyboard to manually scroll through each frame and see the results of your work. You can also render the sequence by following the instructions in Chapter 11.*

7. Add *zoom960.psd* to the Timeline. The still appears in the Composition window.

The pixel dimensions of this image are twice the size of a frame of video. Unlike Final Cut Pro, which automatically scales a still image to fit a video frame, After Effects displays the image at full size, so parts of the image do not appear on screen.

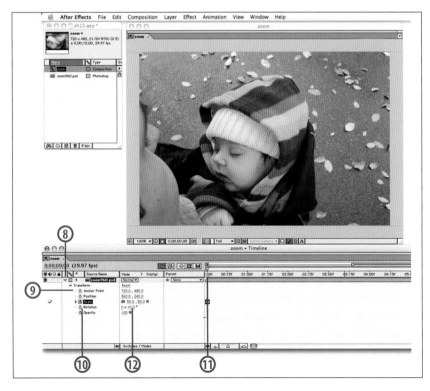

8. Click the twirl-down icon for the first layer in the Timeline.

9. Click the Transform twirl-down menu to access the Transform controls for the layer. (For more detail on the Transform controls and scaling a layer in After Effects, see Chapter 11.)

10. Click the stopwatch icon next to the Scale controls in the Timeline. When you click the stopwatch icon, After Effects enables you to change a property over time. In this case, you will scale the clip to make it smaller in the first frame, and then gradually increase the image to full size, simulating a zoom.

11. Drag the Current Time Indicator to the first frame in the Timeline.

12. Double-click in the first Scale field, and enter 50.

13. Press Return (Mac) or Enter (PC). After Effects adds a keyframe at the Current Time Indicator's present location, and scales the first frame of the image to 50 percent so it fits entirely inside the frame.

14. Drag the Current Time Indicator to the last frame in the Timeline.

15. Double-click the first Scale field, enter 100, and press Return (Mac) or Enter (PC). After Effects adds a keyframe in the last frame of the Timeline, and then scales the last frame of the clip to 100 percent.

16. Preview your effect. As the Composition plays, the still increases in size, until it becomes a tight close-up at the end of the Timeline.

17. Save your work.

Simulating a camera movement

When you changed the size of the image in the last exercise, you also radically changed the framing. When the image appeared in the Timeline at 50 percent of its original size, the entire frame was visible—including a sliver of Luca's father and a nice buffer of empty space. At 100 percent, Luca's face appears too close to the left edge of the frame (in film terms, he's crowding the edge) and there's an unbalanced amount of space at the right edge.

You can compensate for this by repositioning the still. In Chapter 12, you repositioned an entire clip after scaling it to frame out some problematic trees. In this exercise, because you've already added keyframes at the beginning and end of the sequence, you can reposition the still in the last frame of the Timeline without changing the position of the still at the beginning. As a result, the still will gradually change position over the course of the sequence. By combining the camera movement with the zoom you added in the last exercise, you can create a very polished effect.

In Final Cut Pro

Here are the steps for simulating a camera movement in Final Cut Pro:

1. Click the "home" key to ensure the Playhead is in the first frame of the Timeline.

2. Click the Insert/Delete Keyframe button next to the Center field in the Viewer. Final Cut Pro adds a keyframe for the Center value at the start of the sequence. You can now reposition the clip at the end of the sequence without changing its position in the beginning.

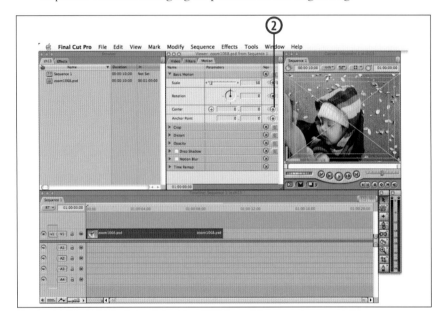

NOTE

You can use the same technique of adding keyframes and repositioning a clip to create simple pans and tilts or complex orchestrated movements in any clip. (Remember, a pan is always a horizontal movement and a tilt is always vertical.) To create a basic camera movement, add one keyframe at each end of the Timeline. To create more complicated pan/tilt combinations, simply add additional keyframes at various points in your sequence and reposition the clip at each point.

3. Click the End key on your keyboard to ensure the Playhead is in the last frame of the sequence.

4. Click the Insert/Delete Keyframe button to add a keyframe in the last frame of the sequence.

5. Click in the Canvas window, and drag the image to reposition it. In the example, I dragged up and to the right to create a greater feeling of balance. Note that as you drag the image, the values in the Center fields change to reflect your adjustments. (To ensure your Browser window looks like the illustration, select View→Image+Wireframe and also select View→Show Title Safe.)

6. Preview the effect. Your sequence now contains a zoom, along with a pan/tilt combination. If you were using a camera mounted on a tripod, this combination would be very difficult to achieve—using keyframes in Final Cut Pro, it's a very straightforward procedure.

7. Save your work.

In After Effects

Here are the steps for simulating a camera movement in After Effects:

1. Drag the Current Time Indicator to the first frame of the Composition.

2. Click the stopwatch icon next to the Position control. After Effects adds a keyframe to the first frame in the Timeline.

3. Drag the Current Time Indicator to the last frame in the Timeline.

4. Click in the Composition window, and drag the image to reposition it. In the example, I dragged up and to the right to create a greater feeling of balance. Note that as you drag the image, the values in the Position fields change to reflect your adjustments. Because you clicked the stopwatch icon in step 2, After Effects automatically adds a keyframe to the last frame of the Timeline when you reposition the image. As a result, changes you make to the last frame of the Timeline don't change the position of the image in the first frame of the Timeline.

5. Preview your Composition, which now contains a zoom, along with a pan/tilt combination.

6. Save your work.

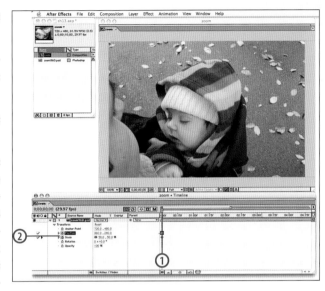

NOTE

You can preview an effect manually by clicking on the Current Time Indicator and dragging it through the Timeline. As you move the Current Time Indicator to the right, your effect plays in the Composition window. If you move the Current Time Indicator to the left in the Timeline, your effect plays in reverse.

You can control the playback of your preview by dragging the Current Time Indicator at different speeds. You can also drag it back and forth through a short section of the Timeline, a technique called scrubbing, to carefully review a particular part of an effect.

Now that you've mastered a few fundamental techniques, try some experiments. Create a new Composition using multiple keyframes to create complicated movements and even more nuanced zooms. If you're looking for inspiration, watch *The Kid Stays in the Picture* by Nanette Burstein and Brett Morgen (it makes a truly innovative use of still images; see Chapter 11) and then build on the techniques you've learned so far.

SIDEBAR

The Kitchen Conqueror and the Power of Still Images

In January of 2004, I went to a conference on digital technology for independent producers, sponsored by the Independent Television Service. (ITVS funds a number of independent projects each year, if you watch public television you've probably seen their work. For more information, take a look at *www.itvs.org.*) The world of independent media is small enough that, at the conference, I ran into an old friend from film school, along with several other people I'd met at similar conferences in years past. For me, one of the main benefits of going to a conference is meeting people who are working in areas of digital media similar to my own, and exchanging ideas. At this conference I met a number of interesting people, including a media designer who had recently earned an MFA from Art Center College of Design in Pasadena and produced a short film as her thesis project that made an innovative and effective use of still images combined with live action video.

In Shereen Abdul-Baki's film, *The Kitchen Conqueror,* the main character walks through the isles of an ethnic food market and slips into daydreams about "marriage, love, independence, and cultural identity." The conscious, tangible world of the main character is shown to the audience as live action video, while Abdul-Baki uses still images in various states of motion to explore her character's dreams and reflections. Abdul-Baki points to a number of creative influences, including Chris Marker's *La Jetée,* a story told entirely with black-and-white still images, voice over narration, and some sound effects (the film was later remade as *12 Monkeys,* starring Brad Pitt and Bruce Willis).

In an email exchange related to the material in this chapter, Abdul-Baki explained that she wanted to explore the concept of Design Cinema (a term she coined for a form combining aspects of live action video, animation, motion graphics, and graphic design) in which visual design was not only an aesthetic consideration but a driving force to communicate what her story was about. Citing Iranian films, which "concentrate on slow, long shots and create a space in which we, the audience, are given time to breathe," she said she also wanted to use stills to create a slower-paced contemplative environment. Likening her work to the international "slow food movement" (in which food is prepared with care and eaten accordingly, as opposed to fast food, which is produced and consumed at high speed and in mass quantities—see *www.slowfood. com*), she said she wanted to define a unique space within Design Cinema "where we can let things stew, enjoy the moments, and allow people to think."

In response to an email in which I asked how she went about integrating stills into her work and animating them, she wrote the following:

> I spent a lot of time sketching on paper and in Photoshop before I moved into motion. I created posters, short animated sequences, and collages. I then created many Photoshop files, in the aspect ratio of the film screen, to sketch out what I wanted to happen visually. When I found that the style and character of the scene was developing, I turned these sketches into storyboards, to visualize the flow in more detail. I was constantly printing out my sketches, storyboards, and written ideas, and plastered them all over the walls around me, so that I could see what I was doing as I was working.
>
> The transitions between the market scenes and the dream sequences were born out of a total frustration with the computer and how the tools were dictating the visual style of my piece, which I was not satisfied with. I left the computer and ran to the Xerox machine, where several hundred copies and photographs later, I ended up with transitions that were far more interesting and which spoke a much simpler and poetic language. I think that it is important to get out of the computer, to try and find inventive ways of approaching things. We are trained, just by the nature of what we see every day, to accept a certain language of visual communication. The nature of digital production lends itself to constant editing, tweaking, and reworking. You can do all sorts of interesting things with just a digital camera.

It's very interesting, if not slightly ironic, that even though she was working with digital video and ultimately mastered her project to DVD, Abdul-Baki made a creative discovery using the analog technology of a photocopy machine. As I wrote in the opening of this chapter, you can use stills in extraordinarily creative ways—perhaps even ways no one has tried before. Most importantly, you can tailor and fine-tune the presentation of each still so it perfectly fits the tone of your film and your artistic sensibility.

You can use the techniques described in this chapter to move a still across the screen (horizontally, vertically, or in any combination you'd like). You can also zoom in or out, and combine zooms with simulated pans and tilts to achieve some very sophisticated effects.

Chapter 14 explores the creation of static and animated titles using Final Cut Pro and After Effects. It also examines ways you can use Photoshop to manipulate still images (including using layer masks to create transparency) and incorporate them into your titles. Chapter 15 then shows you how to bring your title sequence into your project and use keyframe and opacity controls to finesse some very smooth transitions.

You never have to use a static, or nonmoving, image again—unless you'd like to.

Creating Titles, Static and Animated

14

There's an episode of *The Brady Bunch* where the kids make a short film reenacting the Pilgrims' landing at Plymouth Rock. To create the title and credits, they draw on sheets of paper in Magic Marker and then hold each sheet in front of the camera. I saw the episode when I was in perhaps first or second grade and remember thinking the technique was very inventive —why use simple unadorned text when you can draw? As the years have gone by, my aesthetic approach has matured, at least to a certain extent. I no longer think of Bobby and Jan as cinematic innovators, but I still believe that going beyond simple white text on a black background (the default option of many title creation programs) can really benefit your work.

Good titles and credits tell the audience that you care about every detail in your film. A well-designed title sequence engages your audience the moment it appears on screen, and it allows you to set the tone for your viewer's experience. Today's digital editing systems offer fantastic title creation tools, and when you combine them with image-editing programs such as Illustrator and Photoshop, they offer tremendous possibilities to you as a filmmaker. Whether you're making a straightforward documentary or the next *Star Wars* (which has what may easily be the most recognizable and most frequently imitated title sequence of all time), carefully designing your title, and effectively implementing it, are essential steps in the production of your film.

I'm not saying every title you create should be an overproduced technological marvel. *Napoleon Dynamite*, the very funny story of a high school loser who comes into his own, uses low-tech school supplies and even a cafeteria lunch to create one of the greatest title sequences I've ever seen (apparently I'm still impressed by people who make titles by hand and then place them in front of the camera).

Creating Static Titles

There are two kinds of titles, *static* and *animated*, both of which are addressed in this chapter. Static titles don't move. They consist of text (sometimes combined with images) in stationary positions on the screen. Static titles can be faded in and out, and they can be combined with motion elements such an animated background, but by definition, a static title is not animated.

SIDEBAR

Bitmap Versus Vector

Computer graphics fall into two broad categories: bitmap and vector. Bitmap graphics, which are also sometimes called raster images, control pixels individually and enable computers to produce detailed images, such as photographs. Bitmap images can also display very fine gradations of color, so they're often used to create images containing drop shadows. Vector graphics control groups of pixels together, using mathematical calculations called vectors, and are used to create images containing illustrations and text.

Vector graphics are easily scaled and can be increased in size without any loss in quality. Bitmap graphics often don't scale as well, and if a bitmap graphic is scaled beyond 100 percent, it can break down or become pixilated—the pixels become larger and the image starts to look like a blocky mosaic (you may have noticed this if you've tried to enlarge an image in Photoshop). The same characteristics hold true for text. Bitmapped text can break down when it's scaled, while vector-based text can be scaled and fine-tuned without any loss in quality.

Vector-based design programs, such as Adobe Illustrator, are often used by title designers because of the flexibility they afford in text creation and editing. Some designers create titles in a vector-based program, and then import them into After Effects or Final Cut Pro to animate them using keyframes. (For more information on keyframe animation techniques, see Chapter 13.)

Adobe Photoshop is a bitmap editing program, and is often used by title designers incorporating still images into their work. Photoshop also enables designers to manipulate an image's alpha channel to create areas of transparency (see "Creating transparency with a layer mask" in this chapter).

Creating static titles in Final Cut Pro

In the days before desktop video production software, such as Final Cut Pro and After Effects, making even a simple black and white title was a tough business. Creating broadcast-quality video titles required a specialized piece of expensive equipment called a Chyron generator and a highly skilled Chyron operator to run it. In fact, there was a time when Chyron operators made a good living because they were the only ones who knew how to make good, professional quality titles. (The Video Toaster, a hardware-based title generator, offered a more affordable title creation solution and was quickly adopted by public access stations and producers on a budget.) These days, if you're using Final Cut Pro or After Effects, you have sophisticated title creation ability at your fingertips.

This exercise shows you how to make a static title in Final Cut Pro.

1. To access Final Cut Pro's title generator, click the Generators pop-up selector button in the Viewer. A pop-up menu opens.

2. Select Text→Text. The words SAMPLE TEXT appear in the Viewer window.

3. Click the Controls tab at the top of the Viewer window. The text generator's controls become visible.

4. Click inside the Text field, and then using your keyboard, type the name of your title. In this case, I've typed "Big Luca," which is the name of the short film I made for this book.

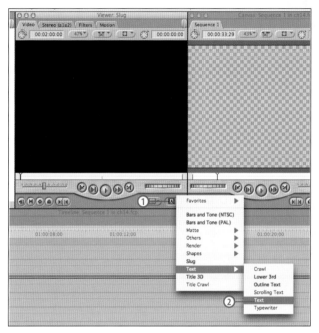

The text generator operates like a word processor. Whatever you type on the keyboard appears in the text generator's controls. If you make a mistake, you can simply press delete, and retype. You can also change the font and point size just as easily as you would in Microsoft Word.

To change the font, size, and color:

1. Click on the font pull-down menu, and select a new font that suits the emotional feel of your work. When you release the mouse, the text generator displays your changes.

2. Drag the Size slider to the left or right to select a new size for your title. Dragging to the left makes the title smaller, dragging to the right makes it larger. The change takes effect when you release the mouse. (You can also enter a size directly in the size field and press return.)

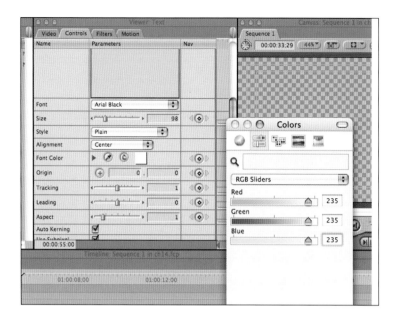

3. Scroll down to reveal the Font Color controls.

4. Click the color sample. The Colors window opens.

5. Using the sliders in the Colors window, select a color for your text. In this example, the title is set to a broadcast-safe white (for more details, see Chapter 11). You can change the color by entering new values in the red, green, and blue color fields (see Chapter 11).

6. When you're satisfied with your title, click the Video tab at the top of the Viewer window. Final Cut Pro closes the text generator controls and creates a new video clip that displays in the Viewer.

Adding your title to the Timeline and adjusting the position

When you create your title, Final Cut Pro doesn't automatically add it to the Timeline. To use your newly created title in a film, you need to place it in a sequence.

1. Drag your title from the Viewer into the Timeline. The title now appears in the Timeline, Viewer, and Canvas windows. By default, Final Cut Pro centers a title in the middle of the frame. If you like your title in the middle of the frame, you're all set; if not, you can easily reposition it.

2. Click in the Canvas window, and drag your title to a new position. When you release the mouse, your changes take effect. In the example, I dragged the title down and to the left, so it fits just inside the title-safe area. (For more detail on the title-safe area and repositioning clips, see Chapter 12.)

3. Save your work.

— NOTE —

Final Cut Pro creates titles on a transparent background, which means you can easily layer your title over a video clip to create a composite sequence. (For more on composite sequences, layering techniques, and stacking order, see Chapter 11.)

Editing your Final Cut Pro title

If you decide at some point that you're not satisfied with your title, you can easily make changes.

1. Double-click your title in the Timeline. The title opens in the Viewer window.

2. Click the Controls tab in the Viewer. Final Cut Pro displays the text generator controls you used to create the title originally, and you can now make adjustments as needed. Instead of creating an entirely new title, Final Cut Pro simply updates the title in the Timeline to reflect your changes.

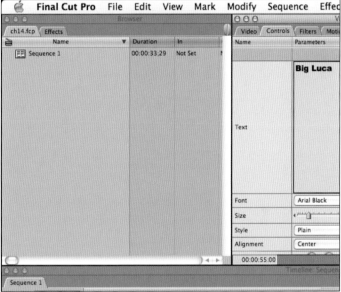

Creating static titles in After Effects

To create static titles in After Effects:

1. Create a new Composition (if you'd like to review how to do this, see Chapter 11).

2. Click the Horizontal Type Tool in the Tools palette.

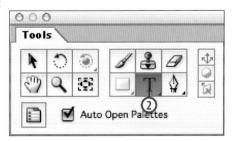

3. Click in the Composition window at the approximate place where you'd like to add your text (don't worry about clicking in the exact spot, you can always reposition your title later).

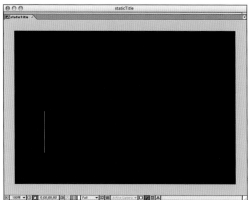

4. Using your keyboard, type the name of your title. The text you type appears in the Composition window at the point you clicked on in the last step. (As in the Final Cut Pro example, I typed "Big Luca," which is the short film I created for this book.) Using the Character palette, you can now adjust the font and point size as easily as if you were using a word processor.

NOTE

Unlike creating a title in Final Cut Pro, After Effects automatically adds your title to the Timeline as you create it.

5. Click the Selection tool in the Tools palette. After Effects automatically selects the text you typed in step 4.

6. Click the Font pull-down menu in the Character palette.

7. Select a new font from the list of available choices. After Effects updates your title in the Composition window.

8. Click the Font Size pull-down menu in the Character palette.

9. Select a new font size from the list. When you release the mouse, After Effects updates your title. (You can also change the font size by clicking in the Font Size field and typing a new value on your keyboard.)

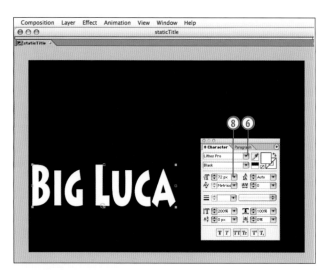

Editing the fill and stroke colors in After Effects

There are two parts to each character in an After Effects title: a stroke and a fill (Figure 14-1). A stroke is the outline of each character—for example, the shape that makes a capital B. The fill is the color inside the outline.

A character can have a stroke and a fill of the same color, or different colors. It's also possible, and sometimes even desirable, to create a character with a stroke and no fill (try this out, it can look really cool). Whatever direction you decide to go in, After Effects makes editing the fill and stroke colors a very straightforward process.

1. Using the Selection tool, click the title to ensure it's selected.

2. Click the Stroke color box. The Color Picker opens.

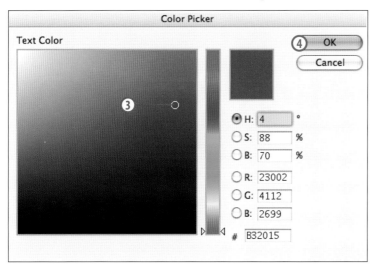

Figure 14-1. In this screen capture of the Character palette, the fill color is set to a broadcast-safe white, and the stroke is set to transparent.

3. Click to select a color for the title's stroke. When you select a color, After Effects displays the color you've chosen in a smaller field.

4. When you're satisfied with the stroke color you've selected, click OK. After Effects closes the Color Picker window, and applies the stroke color you selected.

5. Click the Stroke Width pull-down menu to select a stroke weight for your title (the weight of a stroke is the width of the stoke itself, measured in pixels). In this example, the stroke is set to three pixels. As you've probably guessed, a wider stroke creates a thicker outline.

6. Save your work.

Line Width and the Importance of an NTSC Monitor

Wider lines often display better on video monitors. As a result, small text and text with thin stroke weights will often flicker when displayed on a television screen. (The flicker results from the interlacing of two fields of video to create each frame; see Chapter 2.)

When you're working with titles, it becomes especially important to check your work on a dedicated NTSC monitor—your titles may look very different on a video monitor than they do on a computer screen. If you find that an element in your title appears to flicker, try making the stroke or even the type itself larger. Flickering text can distract your audience and take them away from the content of your film.

For details on connecting an NTSC monitor to your editing system, and why it's a good idea, see Chapter 10. For details on how you can ensure the color in your text is broadcast safe, and will appear the way you want, see Chapter 16.

Resizing and repositioning a static title in After Effects

One of the main benefits of creating titles in After Effects is you can easily change a title at any time. Simply click on the text using the Selection tool, and change any part of the title you think could stand some improvement; this way, you never have to accept a title you're not fully satisfied with. If only everything else in life were that easy.

This exercise shows you how to change the size and position of a title.

1. Using the Selection tool, click the title to ensure it's selected.

2. Click in the Font Size field, and enter 85. (The largest size in the pull-down menu is 72, but you can enter larger sizes directly into the Font Size field.)

3. Hit return on you keyboard (Mac) or Enter (PC). After Effects scales the title to a larger size.

The newly resized title appears in the Composition window; however, it's too big to fit entirely within the frame at the current location. The solution is to move your title to a new part of the Composition window.

When positioning a title, it becomes especially important to work with the title-safe and action-safe guidelines (see Chapter 12 for more details). After all, if the audience can't see your title, what's the point? After Effects does not, by default, display title- and action-safe guidelines. You can, however, easily set them to display in the Composition window.

1. Click the Title-Action safe icon at the bottom of the Composition window. Title-safe and action-safe guidelines now appear in the Composition window.

2. Drag the title to a location inside the title-safe guide. In the illustration, I dragged the title just inside the lower-left corner of the title-safe area. (I also moved the Character palette out of the way so the title would be fully visible in the screen capture.)

3. Save your work.

Animating a Title in After Effects

At this point, you've created and refined a static title in After Effects. You can manually create an animation by keyframing the scale and position controls (see Chapter 13), or you can add one of After Effects' preconfigured title presets. These preconfigured text animations enable you to easily add some very impressive effects to your work. This exercise shows you how.

1. Select Window→Effects & Presets. The Effects & Presets window opens.

2. Click the Animation Presets twirl-down control. The *Text* folder appears.

3. Click the *Text* folder's twirl-down control. A categorized list of subfolders opens, each containing an assortment of preconfigured text effects.

4. Select an effect and drag it onto the layer your title occupies in the Timeline. After Effects applies the preconfigured effect to the title you created in the previous exercises. Your static title is now animated.

5. Preview your effect in the Composition window. If you don't like the effect, select Edit→Undo Apply Animation Preset, and choose another.

6. When you're satisfied, save your work.

NOTE

If you want to experiment, you can drag multiple effects into the Timeline to combine animation presets and create even more complicated configurations. After Effects will automatically combine the presets to create new effects.

SIDEBAR

Choosing the Right Effect for Your Project

A title sequence sets the tone for your project. As a result, the title you create should reflect the content and emotional weight of your film. If you have a somber historical documentary, this may not be the time to experiment with flashy title effects. Likewise, if you're making a wild comedy set on a college campus, you might not want to make a staid white on black title that reminds people of the PBS series *Antiques Roadshow*.

For the example, I selected an animation preset called

Scale Up (it's in the Scale subfolder). *Big Luca* is all about scale and proportion, so I thought an effect that grows text to its full size would be appropriate.

At the end of the day, no matter how cool any effect might look, it carries significantly more impact if it suits the emotional tone of the rest of your work. Depending on the tone of the film you're making, the most appropriate title might be completely static without any effects at all.

Creating Scrolling or Crawling Titles in Final Cut Pro

NOTE

The more contrast you have between the text and the background, the easier your title is to read. The two easiest-to-read text/background combinations are white text on a black background, which is called negative text, and black text on a white background, which is called positive text.

As you create your titles, keep in mind that the darkest possible black and brightest possible white are not NTSC color safe. For more detail on managing the color in your video, see Chapter 11 and Chapter 16.

At the end of just about every movie and television show, the credits roll up from the bottom of the screen and exit through the top of the frame. Part of the reason filmmakers roll credits is they've become a convention—when audiences see the credits roll, everybody knows the show is over. But rolling credits also fill a practical need: they enable a filmmaker to acknowledge everyone who worked on a program in a relatively short amount of time. If people's names were displayed on screen as a series of individual title images, often referred to as title cards (also sometimes called *slates*), it would take far longer to credit the people who worked on a production.

Text that rolls up from the bottom of the screen is generally referred to as *scrolling text*. When text moves across the screen horizontally, it's generally referred to as *crawling text*. In recent years, crawling text has become increasingly common, not so much for use in titles but to announce breaking news stories or severe weather alerts on television.

Like static titles, creating scrolls and crawls once required specialized equipment and highly trained personnel. Now, using Final Cut Pro's title generator, you can create scrolling or crawling titles just as easily as you created a static title earlier in this chapter.

Creating a scrolling title in Final Cut Pro

To create a scrolling title in Final Cut Pro:

1. To access Final Cut Pro's title generator, click the Generators pop-up selector button in the Viewer. A pop-up menu opens.

2. Select Text→Scrolling Text. A blank video clip opens in the Viewer.

3. Click the Controls tab in the Viewer. Final Cut Pro displays the scrolling text generator's controls, which enable you to add text, set the font, point size, alignment, and color. The controls also allow you specify the direction in which your text will scroll.

4. Click in the Text field and type the credits for your film. In the example, I placed an extra line of space between each credit to make the list more readable.

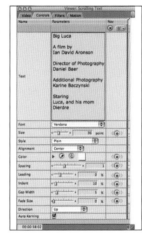

5. Select a typeface from the Font pull-down menu. In the example, I chose Verdana because it's very easy to read. (The fonts Verdana and Georgia were both designed to be legible on screen. Verdana is a sans-serif font, and Georgia is a serif font. Serifs are short lines at the ends of letterforms— for example, small details at the bottom of a letter that are often referred to as "feet.")

M M

Figure 14-2. The letter M (left) is written in Verdana, a sans-serif font, and in Georgia (right), which is a serif font.

6. Select a Size by dragging the Size slider, or by clicking in the Size field and entering a number using your keyboard.

7. Choose an alignment from the Alignment pull-down menu. By default, Final Cut Pro aligns your scroll to the center of the screen. You can choose to align your credits to the left or right, depending on your visual style.

8. Select a color for your credits. You can click the color sample and choose a color from the pop-up menu that opens, or you can click the eyedropper and sample a color from somewhere on the screen.

9. Decide if you want your text to scroll up from the bottom of the screen and out through the top of the frame, or if

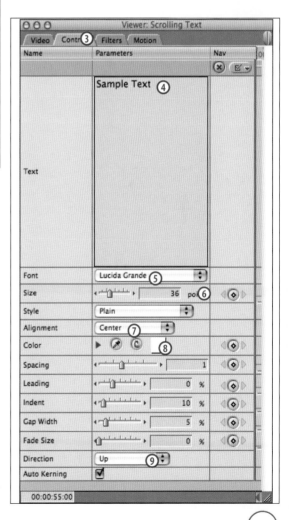

NOTE

By default, Final Cut Pro scrolls text up from the bottom of the screen. To change the direction, click the Direction pull-down menu, and select Down.

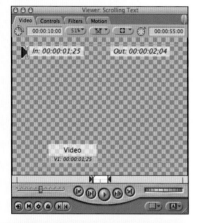

NOTE

You can composite your scrolling title onto another layer of images by dragging your title from the Viewer onto a video track above another clip. See Chapter 11.

Creating scrolling text requires a fair amount of processing power from your computer, so more than likely, you'll need to render the sequence before you can play it back at full speed. For details on rendering and previewing effects, see Chapter 11.

you want your text to scroll down from the top of the frame and out through the bottom. Audiences are used to seeing credits scroll up from the bottom of the screen, so that's what they'll expect when they see your film. If you'd like to surprise them you can try scrolling your titles in from the top, but some people in your audience might not have an entirely positive reaction. I recently showed a class how to scroll text down from the top of the screen and when we previewed the results I found it physically disconcerting.

10. When you're satisfied with your adjustments, click the Video tab in the Viewer to display the title you just created. The scrolling title now appears in the Viewer window. (Because the title scrolls in from the bottom of the screen, the first frame of the title that displays in the Viewer will be blank.)

11. Press Play in the Viewer to preview your title. When you're working with credits, it's especially important to ensure all your text remains within the title-safe boundary. If any of your text falls outside the title-safe area, click on the Controls tab and change the font size.

12. Save your work.

Adding your credits to the Timeline

Before you can use your scrolling title in your film, you need to add it to the Timeline. This process works the same way as adding the static title you created earlier in this chapter.

1. Drag your title from the Viewer into the Timeline. The title now appears in the Timeline, Viewer, and Canvas windows.

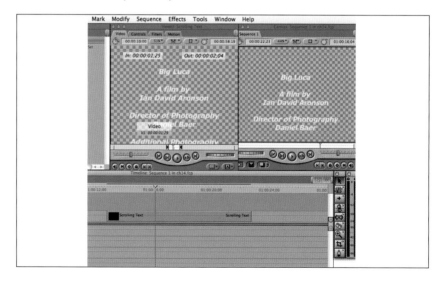

2. Save your work.

Creating a crawling title in Final Cut Pro

Networks such as CNN, Fox News, and especially CNBC use crawls (text that moves across the screen horizontally) so often they're starting to become commonplace. (Interestingly, *The Teletubbies* uses a crawl for their credits—the show can be surprisingly avant-garde.)

Crawling text can really distract an audience from the content of your show, so this technique is something you may want to use sparingly.

1. Click the Generators pop-up selector button in the Viewer, and select Text→Crawling Text. A blank crawling text clip opens in the Viewer.

2. Click the Controls tab to display the crawling text controls.

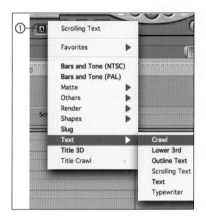

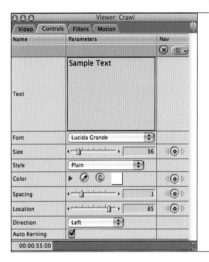
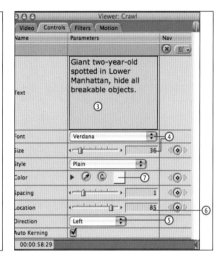

3. Enter your text in the Text field.

4. Select a font and a size.

5. Select a direction from the Direction pull-down menu. Left, which is the default, means your text will crawl onto the screen from the right side and out through the left side. If you're creating text in English, or any other language that reads a sentence from left to right, this is pretty much the only functional option. If you're creating a project for a Hebrew-reading audience, or another audience that reads from right to left, choose Right from the pull-down menu.

6. Set the vertical position of the text using the Location control. A lower number in the Location field positions the text closer to the bottom of the screen; a higher number positions the text toward the top. While people generally put crawling text at the bottom of the screen, you can put the text anywhere you'd like—it's your film.

7. Select a color for your text.

--- NOTE ---

You can edit a scrolling or crawling title in Final Cut Pro at any time by clicking on the title where it appears in the Timeline, and then clicking the Controls tab in the Viewer. The changes you make will automatically update in the Timeline and Canvas windows.

Creating scrolling text requires a fair amount of processing power from your computer, so more than likely, you'll need to render the sequence before you can play it back at full speed. For details on rendering and previewing effects, see Chapter 11.

8. When you're satisfied with your adjustments, click the Video tab to display at the title you just created. The crawling title now appears in the Viewer window. (Because the title crawls in from off screen, the first frame of the title that displays in the Viewer will be blank.)

9. Press Play in the Viewer to preview your crawl and to make sure everything is title-safe.

10. When you're ready, add your crawling title to a sequence by dragging it from the Viewer window into the Timeline (the process is the same as adding a scrolling title, described earlier in this chapter).

11. Preview the sequence, which now contains your crawl.

12. Save your work.

Creating Subtitles and Lower Third IDs

You can use the static title creation techniques described in this chapter to subtitle your film, or to create lower third and various other forms of onscreen text.

If you're creating something for an international audience, or if your project has dialog in more than one language, onscreen text becomes particularly important. Subtitles enable you to bring the full detail of your story to audiences, without dubbing actors' voices or making other changes that feel like an artistic sacrifice. (I once had a job dubbing Hollywood movies and TV shows into Spanish—the end product was always strikingly different from the original.)

Lower third IDs, also referred to as lower thirds, are short lines of text that filmmakers use to identify people or locations that appear onscreen. When you're watching a documentary and someone appears onscreen for the first time, his appearance is often accompanied by a lower third to explain who he is and why he's in the film. Another example of a lower third is when you watch your local news and the reporter's name appears at the bottom of the screen as she introduces her story. Filmmaker Bill Jersey (also mentioned in Chapter 13) developed the "20-minute rule," for lower third IDs. His 20 minute rule was that people should get a lower third the first time they appear on screen, and again if they reappear after 20 minutes or more. You may decide to use lower thirds more or less frequently in your film—it's up to you to decide what works best.

Although lower thirds can be very helpful, if not essential, to explaining who someone is and putting their appearance in context, if overused, they become redundant. As a result, you don't need to add one every time the same person appears in your film.

Similarly, if you add overly elaborate text to your screen, your audience may not be able to pay attention to the text and your story at the same time. When you're adding text to the screen, it's really important to do it in a way that doesn't take the audience out of your film. Making good onscreen text is clearly important, but if you overload your audience with information, they may stop paying attention.

Research has shown that people watching images and text on screen process both as visual information. People can process only a limited amount of information at any one time, and if you provide too much visual information, your audience will have to choose some things to pay attention to and others to ignore. If you show text on the screen while people are speaking, it becomes especially hard to process both at the same time, and people may start to tune things out. If the audience begins to ignore parts of your story, your film loses potency.

Adding Still Images to Your Titles

As I mentioned earlier, After Effects and Final Cut Pro both create titles on a transparent background so you can composite them into a sequence containing other visual elements. For example in Chapter 5, I wrote about the title design for *Six Feet Under*, in which the designers composited the titles into thematically appropriate frames of video such as a hospital gurney being wheeled down a hallway.

You can also add still images to your static or moving titles. Using an image-manipulation program such as Adobe Photoshop, you can edit your photos so they work with your titles exactly as you'd like them to.

The following exercises show you how to create transparency in a photo using a layer mask, and how to add the photo to your title sequence.

Creating transparency with a layer mask

When you draw a layer mask in Photoshop, part of the image becomes transparent while the rest remains fully opaque (for a review of transparency and alpha channels, see Chapter 11). You can then import your edited image into Final Cut Pro or After Effects, and composite it into a sequence along with your title. Because you can draw a layer mask in any shape you'd like, you can use this technique to edit a still image into a shape that perfectly fits with your title.

1. Open the image in Photoshop. (You can do this by simply double-clicking on the file, or by opening Photoshop, selecting File→Open, and navigating to the file on your hard drive.)

> **NOTE**
>
> *Photoshop works with a series of layers much like Final Cut Pro and After Effects. Working with a layer mask enables you to mask parts of a layer and make them transparent. When you import the masked image into Final Cut Pro or After Effects, the image retains its transparency.*

> **NOTE**
>
> *To follow along with this exercise:*
>
> *Copy the ch14 folder from the Exercises folder on the StartToFinish DVD into your hard drive. The folder contains two Photoshop documents: luca534.psd and luca480.psd. Both are stills of Luca and his father; luca534.psd is sized for use with Final Cut Pro (use this file if you plan to import your edited image into Final Cut Pro); luca480.psd is sized for use with After Effects (use this file if you plan to import your edited image into After Effects). For more on compensating for pixel aspect ratio see Chapter 11.*

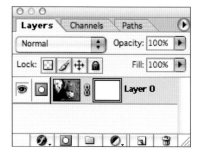

2. Click the Add layer mask button in the Layers palette. (If the Layers palette is not visible, select Windows→Layers.)

When you click, the Add layer mask button Photoshop adds a layer mask to your image. You can now define the shape of the mask using Photoshop's drawing tools. Areas that you draw in black will become transparent. The rest of the image remains fully opaque.

In the following steps, I'm going to draw a layer mask to make everything in the image transparent except for Luca's face. When composited into the title, it will make a very nice effect.

1. Click the Brush tool in Photoshop. The Brush tool enables you to draw on an image by clicking in a location and dragging the mouse. When you added the layer mask in step 2, Photoshop set the foreground color in the Tool palette to black and the background color to white. This means that when you draw with the brush, Photoshop will now draw in black.

2. Click near Luca's face, hold down the mouse, and drag around the outline of his face.

3. Continue drawing until you've isolated Luca's face from the rest of the image.

4. Click the Lasso tool in the Tool palette. You can use the Lasso tool to select very large areas of the image at once.

5. Click in part of the image and drag to draw a large, closed shape. When you release the mouse, Photoshop selects the area of the shape you've drawn. In the following sequence of actions, you can fill the selected

area with black. It serves the same purpose as drawing in black with a brush but enables you to fill large areas of the image more quickly.

6. Select Edit→Fill. The Fill window opens.

7. Select Black from the Use pull-down menu. This determines the color Photoshop uses to fill the selected area. As mentioned earlier, filling the area with black will make it transparent.

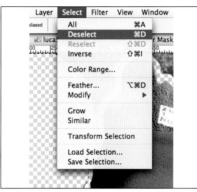

8. Click OK. Photoshop fills the selected area and it becomes transparent.

9. Choose Select→Deselect. Photoshop deselects the area you drew with the Lasso tool in step 6. You can now use the Lasso tool to select another area and make it transparent.

10. Continue to use the Lasso and Brush tools to make everything in the image transparent, except for Luca's face.

11. Save your work.

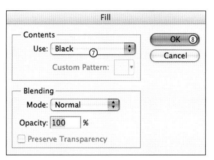

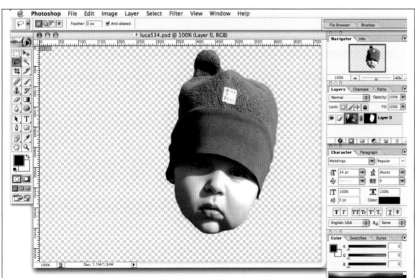

Bringing the image into your title sequence

When you save a Photoshop document containing a layer mask, Photoshop saves the transparency information as part of the document. As a result, you can add the image to your title sequence, and use the transparency to create a very nice composite.

Using your edited still image in After Effects

To insert the image you edited in the last exercise into your After Effects Composition:

1. Import the edited version of *luca480.psd* into After Effects (select File→Import→File).

After Effects opens a window named *luca480.psd* (this window opens whenever you import a layered Photoshop file, and is explained in detail in Chapter 11).

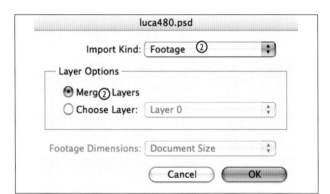

2. Leave the Import Kind pull-down menu set to Footage and the Layer Options radio button set to Merged Layers.

3. Click OK. After Effects imports your file and closes the window.

4. Drag *luca480.psd* from the Project window into the Composition window. After Effects automatically adds the image of Luca to a higher layer in the Timeline. Because Luca's face appears on a transparent background, he only partially obscures the title.

5. Position Luca in the part of the frame where you think he looks best. (You can move an image in the Composition window by clicking and dragging. For more on repositioning a clip, see Chapter 12.)

> **NOTE**
>
> *In the example, I placed Luca in the top right of the screen. Positioning Luca diagonally across from the title creates a nice sense of balance in the frame. You may notice that parts of Luca's face fall outside the action-safe boundary, meaning they'll be cut off by the edge of the screen. I like the way that looks, as well as the fact that his image cuts off the top of the "A" in the title. They both fit the film's visual theme of a boy who's bigger than everything around him.*

6. Select Effect→Perspective→Drop Shadow. After Effects opens the Effect Controls window for *luca480.psd*, which displays the controls for a Drop Shadow effect. Adding a drop shadow lends a nice touch to the 3D effect of the title composite. The image of Luca is clearly on a higher layer than the title, and adding a drop shadow gives a nice bit of depth.

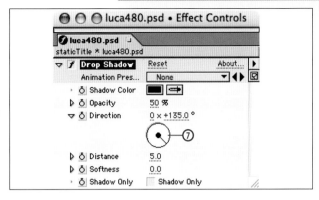

7. Drag the Direction dial clockwise, so the shadow is angled down and to the left of Luca's face. The light in the photo is coming from the top right, and the angle of the photo needs to match for the effect to work.

8. Change the Opacity from 50 percent to 70 percent. This makes the shadow darker and more visible.

9. Change the Distance setting from 5 to 15. This extends the size of the shadow farther away from Luca's face, and again makes the effect more visible.

10. Preview your composite. The title rises up smoothly, until it's partially obscured by Luca's face. The drop shadow adds a nice element of separation between the layers and helps create a very polished effect.

11. Save your work.

Using your edited still with Final Cut Pro

After Effects and Photoshop are both made by the same company, and as I mentioned earlier in the book, they're made to work together. Photoshop and Final Cut Pro are made by different companies, and even though they're compatible, using them together sometimes requires a workaround. Importing an image with a layer mask into Final Cut Pro is one of those times.

Even though you successfully added a layer mask earlier in this chapter, if you import *luca534.psd* into Final Cut Pro with the mask intact, Final Cut may not recognize the transparency information. The trick to making the layer mask work most effectively is to add another layer and then merge your layers together.

1. Open *luca534.psd* in Photoshop.

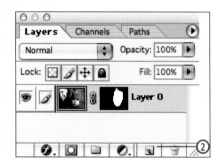

2. Click the Create a new layer icon in the Layers palette (if the Layers palette isn't visible, select Window→Layers). Photoshop adds a new, empty layer to the document.

3. Click the button at the top right of the Layers palette. A pop-up menu opens.

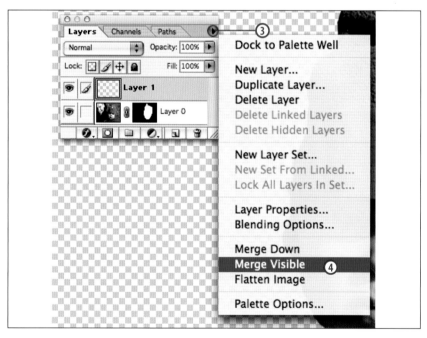

4. Select Merge Visible from the menu.

Photoshop merges the layers, producing a document that has only one. Although the document retains all its transparency information, it no longer contains the layer mask you created. The image is now ready to import into Final Cut Pro.

To add the image to your Final Cut Pro title, do the following:

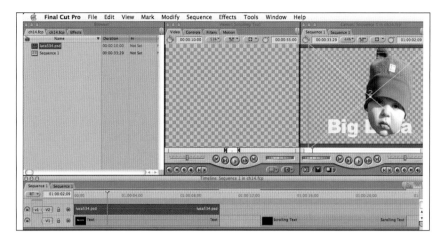

1. Import the reworked *luca534.psd* file into Final Cut Pro.

2. Drag the file into the Timeline above your main title. Final Cut Pro adds a new video track and composites the image of Luca over the title text.

3. Click in the Canvas window and drag Luca to a part of the frame you feel good about. In the example, I dragged Luca to the upper-right corner. As in the After Effects example, I placed Luca diagonally opposite the title text to balance out the frame. I also positioned him so he would be partially framed out by the edge of the screen, adding to the visual theme of a boy who's bigger than his surroundings.

4. Save your work.

This chapter introduced you to a full repertoire of desktop title creation techniques. Working in Final Cut Pro and After Effects, you created static and animated titles, adjusted stroke color and weight, and even created scrolling credits and a crawling breaking news alert. You also added a layer mask to custom tailor an area of transparency in a still image using Photoshop, and then added the image to your title sequence.

Not bad for a day's work.

The next chapter shows you how to bring a completed title sequence into your film, and brings you one big step closer to finishing your project.

Bringing Your Title Sequence into Your Project

15

Now that you've created some sharp-looking titles, it's time to bring them into your project. Rather than simply slapping them into the Timeline, you can exercise complete control over the way each title appears to your audience—including how it enters and leaves the screen. Whether you're working with static or animated titles, you can decide where they appear on screen, whether they pop on at full opacity or gently fade in, and how they interact with other elements of your opening sequence.

This chapter walks you step-by-step through the process of creating the opening sequence of *Big Luca*, using the titles you created in Chapters 12, 13, and 14. The composites you created in the previous three chapters now exist as separate sequences if you're working in Final Cut Pro, or multiple compositions if you're working in After Effects. The following exercises show you how to combine these separate elements into a single, flowing visual montage.

Your goal as a filmmaker and title designer is to draw the audience into your story. How you present your titles and incorporate them into your opening sequence deserves the same amount of care you put into creating them.

Building Your Opening Sequence in Final Cut Pro

This section walks you through laying out your opening sequence, using Final Cut Pro.

Creating a main title sequence

The first step in creating a main title sequence, as you may have guessed, is to create a new sequence. This new sequence enables you to place all your separately created title elements into one single composition, and then add transitions to seamlessly join each element to the next.

1. Open Final Cut Pro. By default, Final Cut Pro opens the last project you worked on. At this point, since you're working with many different elements at the same time, it can be very helpful to close existing projects and open only the project files you need, one at a time, as you're ready to work with each.

2. Select File→Close Project. Final Cut Pro closes the current project that's open on your computer. If you have more than one Final Cut Pro project open on your computer, repeat this step until no project is open.

3. Select File→New Project. Final Cut Pro opens a new project. This new project provides you with a blank slate that you can use to create your final opening sequence.

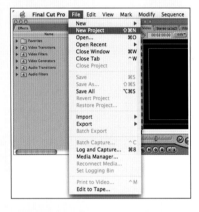

4. Select File→Save Project As.... The Save window opens.

Immediately saving a new project ensures you won't lose any work in the event your system crashes, and it also helps you to keep your project organized—if you can't find your project files when you need them, they're of very little use.

5. Navigate to the location on your computer where you'd like to save the file.

6. Enter a name in the Save As field (in the example, I named the file *ch15. fcp*).

7. Click Save to save the file and close the window.

By default, Final Cut Pro creates a new sequence for each new project file. In this illustration, the sequence is named Sequence 1. Giving the file a more descriptive name will help you stay organized—later in this chapter, you'll have more than one sequence open at the same time, and giving the sequences separate names helps you tell them apart.

8. Click the name of the sequence in the Browser window once to select it.

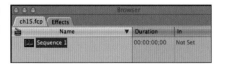

9. Click the name of the sequence again. The sequence name appears in blue. Typing a new name on your keyboard will change the name of the sequence.

10. Name the sequence mainTitle.

11. Press return. Final Cut Pro changes the name of your new, blank sequence.

You're now ready to start adding material and combining your separate title elements into a finished product.

Adding the camera movement and title graphic to your new sequence

Final Cut Pro enables you to open more than one project at the same time. This makes it much easier to copy material from one project into another, and to keep your files organized. Older digital editing programs allowed editors to open only one project at a time, which made sharing the material in a project considerably more difficult.

This exercise shows you how to open two separate composite sequences you created in earlier chapters, and add them to the Timeline of your newly created mainTitle sequence.

1. Open the project containing the simulated camera movement you created in Chapter 13. (You can also open the *ch13.fcp* project file on the DVD that came with this book, just be sure to copy it to your computer first.)

Final Cut Pro opens the project from Chapter 13 and keeps *ch15.fcp* open at the same time. The simulated camera movement appears in the Timeline of Sequence 1. The next step is to move the camera movement from its current location into the Timeline of your blank mainTitle sequence. Believe it or not, this is as simple as cutting and pasting.

2. Click the simulated camera movement clip in the Timeline to ensure it's selected.

3. Select Edit→Copy. Final Cut Pro copies the clip along with the scale and positioning effects you applied to it earlier. This saves you the trouble of repeating all the steps in Chapter 13 when you paste the clip into your new sequence.

---- **NOTE** ----

To complete the exercises in this section, you can use the files you created yourself as you followed along with the instructions in the previous three chapters. If you prefer, you can also use the completed Final Cut Pro project files for Chapters 12 through 14, which are available on the DVD that ships with this book. To access the completed project files:

1. *Navigate to the Exercises folder on the StartToFinish DVD. The folders for Chapters 12 through 14 each contain one Final Cut Pro project file.*

2. *Copy the project files (ch12. fcp, ch13.fcp, and ch14.fcp) to your computer.*

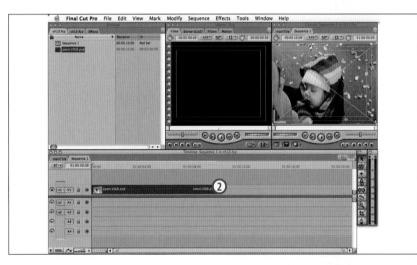

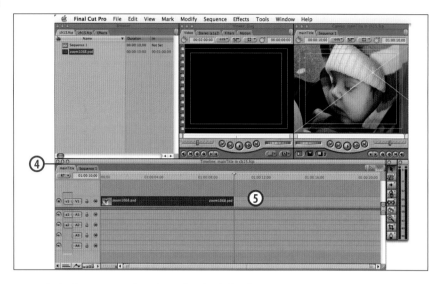

4. Click the mainTitle tab at the far left of the Timeline. This selects main-Title as the active sequence and enables you to paste material into it.

5. Select Edit→Paste. Final Cut Pro adds the clip to the mainTitle sequence, with all your effects intact. (This may actually be the simplest thing you do all day.)

6. Drag your Playhead through the sequence to preview the effect. Just as before, Luca's face grows to fill the frame in a very smooth simulated zoom.

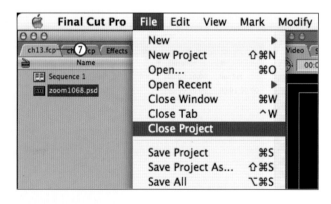

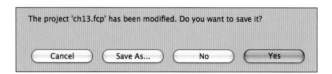

7. Click on the *ch13.fcp* tab in the top left of the Browser window. Now that you've copied the camera movement effect from the Chapter 13 project file, you no longer need to keep the file open. Closing it can help avoid confusion.

8. Select File→Close Project.

Final Cut Pro asks if you want to save your changes to the Chapter 13 project file. Since you didn't actually make any changes to the file—you just copied a piece of it—there's no reason to save any changes.

9. Click No. Final Cut Pro closes the dialog box and saves the Chapter 13 project file without making any changes. This helps preserve the integrity of the original file, in case you'd like to go back and reuse part of it in the future.

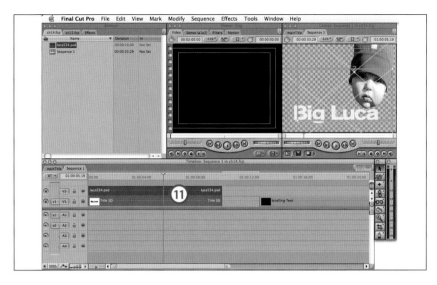

10. Open the project file you created for Chapter 14. (Alternatively, you can use the *ch14.fcp* project file on the DVD that came with this book, just be sure to copy it to your computer first.)

Final Cut Pro opens your main title, which contains the title graphic itself and the edited still image of Luca, in the Timeline.

11. Shift-click both the title graphic, on track V1, and the edited still, on track V2, to ensure they're both selected.

12. Select Edit→Copy. Final Cut copies both clips, along with their positioning information and stacking order. (For more information on positioning a clip, see Chapter 12. For a detailed explanation of stacking order, see Chapter 11.)

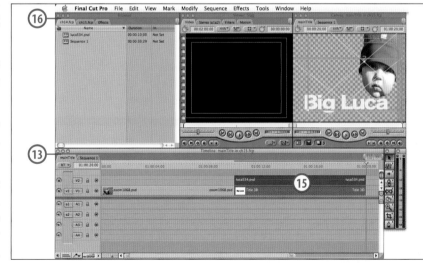

13. Click the mainTitle tab in the Timeline to select it as the active sequence.

14. Place the Playhead just after the end of the *zoom1068.psd* clip in the Timeline.

15. Select Edit→Paste. Final Cut Pro adds both the clips from the Chapter 14 project file to the Timeline and maintains their positioning information and stacking order. Final Cut Pro also adds a new video track to accommodate the edited still image.

16. Click the *ch14.fcp* tab in the top left of the Browser window to ensure the Chapter 14 project is selected.

17. Select File→Close Project and don't save any changes to the Chapter 14 project file. As I mentioned earlier, closing a project file as soon as you're done with it is always a good idea and can help you avoid confusion—especially when you're working with material from many different projects.

18. Drag the Playhead through the Timeline to preview your results. After the simulated camera movement fully zooms in on Luca, the sequence cuts to the main title text and the edited still image. This means that you've successfully woven two separate composites into a single unified sequence.

19. Save your work.

Nice job.

Now you're ready to polish the transition by manipulating the images' opacity to fade seamlessly from one to the next.

Fading your composites off and on the screen

As things stand now, the title image abruptly replaces the gentle zoom. This is a fairly jarring transition, and you might want to use it if you're purposefully trying to unsettle your audience, but it's not your only option. This exercise walks you through the process of easing out of the zoom and into the title sequence. The exercise also shows you how to fade the image of Luca onto the screen after the main title text.

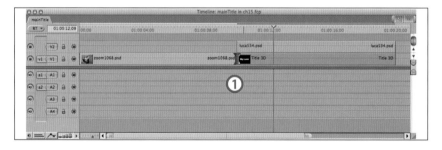

1. Click in track V1 of the Timeline where the end of the simulated camera movement meets the start of the main title text. Final Cut Pro selects the end, or tail, of the first clip and the start, or head, of the second clip.

2. Select Effects→ Video Transitions→Dissolve→Dip to Color Dissolve. Final Cut adds a transition that fades the tail of the first clip to black, and then fades the head of the second clip up from black to become fully visible.

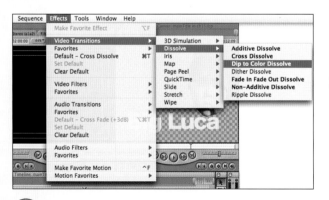

This fade creates a very nice, smooth segue between the clips and matches the gentle pace of the simulated zoom. You can make the effect even more polished by separately fading in the edited still image slightly later in the Timeline.

3. Click the head of the edited still image on track V2 and slide it to the right. (In the illustration, I positioned the head of the clip just after the end of the dip to color dissolve transition on track V1.) Positioning the head of the clip after the end of the

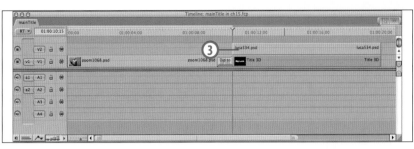

transition effect means the title text will fade up fully before the edited still becomes visible.

4. Use the arrow keys on your keyboard to scroll through the Timeline and preview the sequence. The first clip fades out gently, the second fades in just as gently and then the still image pops onscreen abruptly. To create a smoother effect overall, you can keyframe the still's opacity to fade it in as well.

5. Control-click the still image clip on track V2 of the Timeline, and choose Open in Viewer from the contextual menu that opens. The clip opens in the Viewer window.

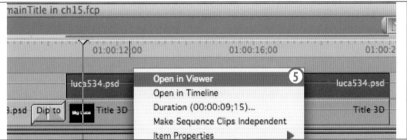

6. Position the Playhead in the first frame of the still image clip.

7. Click the Motion tab in the Viewer. The clip's Motion controls become available.

8. Click the Opacity controls' twirl-down icon to access the opacity settings.

9. Click the Ins/Del Keyframe button in the opacity controls to add a keyframe in the first frame of the still image clip.

10. Drag the clip's Opacity slider to 0. This makes the clip fully invisible. (For a detailed explanation of keyframes, see Chapter 13.)

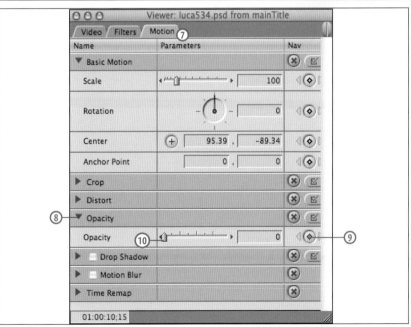

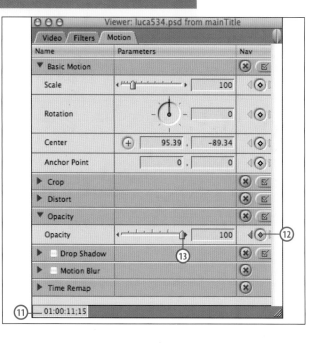

11. Drag the Playhead about one second, or 30 frames, to the right. The timecode window in the bottom-left corner of the Viewer displays the Playhead's current location. (In this illustration, the Playhead is positioned exactly one second later than it was in the previous illustration.)

12. Click the Ins/Del Keyframe button in the opacity controls to add a new keyframe to the still image clip.

13. Drag the Opacity slider to 100. This makes the still image fully visible at the second keyframe.

14. Use the arrow keys to scroll through the Timeline and preview your effect (or render the sequence using the steps described in Chapter 11). The simulated camera movement gently fades out and is replaced by the main title text that gently fades onto the screen, followed shortly by the edited still image of Luca, which now gently fades onscreen as well.

15. Save your work.

SIDEBAR

Nesting Versus Cut and Paste in FCP

When you're working in Final Cut Pro, cutting and pasting is not the only way you can place one sequence inside another. You also have the option of nesting a sequence, which places an entire sequence inside another, not as a series of clips but as one single item.

Once a sequence is nested inside another sequence, you can no longer manipulate each clip individually. You can, however, apply an effect to all the clips in a nested sequence at once. Editors often nest sequences before applying filters such as the Broadcast-Safe color-correction filter. (For more on color correction, see Chapter 16.)

The interviewSelects sequence, pictured above, contains several audio and video clips edited together. When viewed in the Timeline, the different clips are visible inside the sequence.

To nest the interviewSelects sequence inside another sequence:

1. Drag the sequence from the Browser directly into the Viewer window. The sequence opens in the Viewer window as if it were a single clip. (If the same sequence is open in the Timeline, Final Cut Pro automatically removes it from the Timeline when an editor drags it into the Viewer.)

2. Press the Insert Edit button in the Canvas window. Final Cut Pro edits the entire interviewSelects sequence into the Timeline as if it were one continuous clip.

Building Your Opening Sequence in After Effects

Unlike Final Cut Pro, After Effects does not enable an editor to open more than one project at once. Instead, you can import multiple projects into a single After Effects project to combine material from different sources.

This section shows you how to import several After Effects projects into a single new project and combine various compositions to build the opening sequence of your film.

Opening a new master project

To start a new master project in After Effects:

1. Open After Effects. When you open the application, After Effects automatically opens a new, untitled project. This new project will be the master project you use to create your finished title.

2. Select File→Import→Multiple Files.... The Import Multiple Files window opens.

3. Navigate to the first of the After Effects projects you'd like to import.

4. Select the project file and click Open. After Effects imports the file and then re-opens the Import Multiple Files window, enabling you to import additional files.

5. Import the additional project files you'd like to work with, and when you're finished, click Done.

After Effects creates a folder in the Project window for each project you import. In this figure, I imported project files for Chapters 12, 13, and 14. The folders After Effects created for the three files contain the compositions I created in each project, along with any still images and footage the project files contain. (Remember, After Effects doesn't import copies of media files such as still images or video footage, it simply creates links to the original files.)

6. Select Composition→New Composition. The Composition Settings window opens.

NOTE

To complete the exercises in this section, you can use the files you created yourself as you followed along with the instructions in the previous three chapters. If you prefer, you can use the completed After Effects project files for Chapters 12 through 14, which are available on the DVD that ships with this book. To access the completed project files:

1. *Navigate to the Exercises folder on the StartTo-Finish DVD. The folders for Chapters 12 through 14 each contain one After Effects project file.*

2. *Copy the project files (ch12.aep, ch13.aep, and ch14.aep) to your computer.*

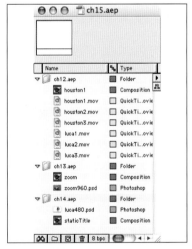

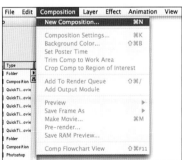

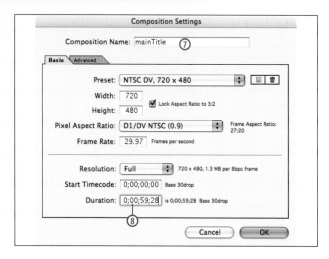

7. In the Composition Name field, enter mainTitle. This gives the composition a descriptive name, which will help you stay organized as the project becomes more complex later on.

8. Enter 0;00;59;28 in the Duration field. This sets the duration of your new composition to just short of one minute. Because you'll be using this composition to combine a series of composites you created earlier, a longer duration provides you with some added flexibility.

9. Click OK. After Effects closes the Composition Settings window and creates your new mainTitle composition.

Adding the camera movement and title animation to your opening sequence

To add the camera movement and animation, do the following:

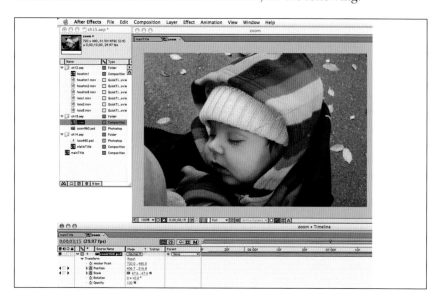

1. Double click the zoom composition in the *ch13* folder to open it in the Timeline.

2. Click once in the Timeline, to ensure the Timeline is the active window.

3. Select Edit→Select All to select the contents of the zoom composition's Timeline.

4. Select Edit→Copy. After Effects copies the image of a sleeping Luca, along with the scale and positioning effects you added in Chapter 13.

5. Click the mainTitle tab in the Timeline to make mainTitle the active composition.

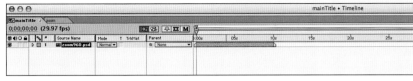

6. Select Edit→Paste. After Effects pastes the still image into your new mainTitle composition, with all the effects intact.

7. Drag the Current Time Indicator through the Timeline to preview your results (or render the composition as described in Chapter 11). The still image slowly fills the frame, ending in a tight close-up of Luca's face.

You've successfully transferred the simulated camera movement from one composition to another. Nice work.

8. Double-click the staticTitle composition in the project window to open it in the Timeline.

9. Click in the Timeline to ensure it's selected as the active window, and then choose Edit→Select All. After Effects selects both the clips in the Timeline of the staticTitle composition.

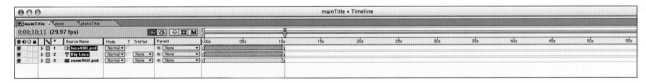

10. Click the mainTitle tab in the Timeline.

11. Select Edit→Paste. After Effects adds the edited still image of Luca's face to the Timeline, along with the title text you created in Chapter 14. By default, After Effects placed both clips at the head of the sequence, obscuring the simulated camera movement. You can easily fix this by dragging the clips to the right.

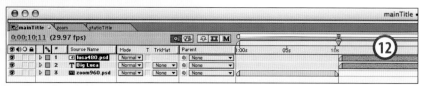

12. Click one of the clips and drag it past the tail of the simulated camera movement on layer 3 (both clips remain selected after you pasted them into the Timeline in step 11, so when you drag one they both move together).

13. Drag the Current Time Indicator through the Timeline to preview the new material you just added. The mainTitle composition now contains the simulated camera movement and the title composite. Shortly after the first clip ends, the second two clips pop onscreen.

You've successfully joined two composite effects into one composition (very impressive work, by the way). Now you can polish the transition between the composites to make your opening sequence even stronger.

Setting the opacity in After Effects to finesse the transition from one composition to another

The way the mainTitle composition is currently put together, the first clip abruptly disappears from view, and after a brief pause, is replaced by the title text and an edited still image. By working with the opacity controls for each clip, you can ease the transition from one composite to the next. This exercise shows you how.

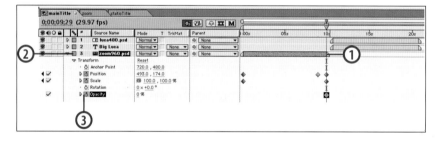

1. Place the Current Position Indicator in the last frame of the simulated camera movement on layer 3, which is the bottom layer.

2. Click the twirl-down icon to access the Transform controls for layer 3, then click the twirl-down icon next to Transform to display the full set of controls associated with the simulated camera movement.

The Position and Scale controls display the settings you added in Chapter 13. In this exercise, you use the Opacity controls to gently fade the screen to black at the end of the zoom.

3. Click the stopwatch icon next to the Opacity controls, and set the Opacity to 0. After Effects adds a keyframe to the last frame of the simulated zoom and sets the opacity of the image to 0, which means fully invisible. (For more detail on using keyframes and adjusting the Transform controls, see Chapter 13.)

4. Move the Current Time Indica-tor about one second to the left. (The timecode field in the top left of the Timeline window displays the

Current Time Indicator's exact location.)

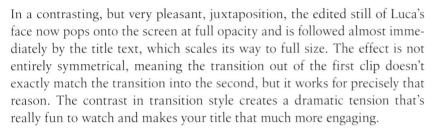

5. Set the opacity to 100. The clip now starts out fully visible, and then gently fades to black over the final 30 frames.

6. Drag the Current Time Indicator through the Timeline to preview your composition, or you can render it. The simulated zoom fades off the screen nicely, matching the gentle movement of the zoom itself.

In a contrasting, but very pleasant, juxtaposition, the edited still of Luca's face now pops onto the screen at full opacity and is followed almost immediately by the title text, which scales its way to full size. The effect is not entirely symmetrical, meaning the transition out of the first clip doesn't exactly match the transition into the second, but it works for precisely that reason. The contrast in transition style creates a dramatic tension that's really fun to watch and makes your title that much more engaging.

7. Save your work—remember, you can never save too often.

── N O T E ──

If you want, you can adjust the opacity controls to make the edited still image of Luca's face fade onto or off of the screen. You can also leave the composition as is. As the filmmaker and title designer, you're the captain of the ship.

S I D E B A R

Nesting a Composition in After Effects

In the previous exercise you copied material from one composition and pasted it into another. After Effects also enables you to nest a composition inside another, by using a slightly different process.

When opened directly in the Timeline, the staticTitle contains two clips (the title text graphic and the edited still image of Luca's face). Because these images appear as separate clips in the Timeline, you can manipulate each one independently of the other—for example, you can change the opacity of one of the clips in the Timeline while leaving the other undisturbed.

If you create a new composition and drag the staticTitle composition inside the new Timeline (as in the illustration above), the contents of staticTitle appear as one clip and not two. Any changes you apply to staticTitle in the Timeline, such as opacity, position, or duration, will apply to both clips at the same time. In short, because the sequence is now nested inside another composition's Timeline, it functions as one single clip.

Joining Your Opening Sequence to the Body of Your Film

Now that you have a nicely executed opening sequence, it's time to segue into the rest of your film. In addition to the raw footage in the *Ch12* folder on the StartToFinish DVD, which you can use to polish your skills as you design various new effects sequences, the *Ch15* folder contains a number of edited interview clips you can work into your film as well.

The interview clips feature Luca's mother delivering deadpan descriptions of how large her son has suddenly gotten. These low-key interview clips make the over-the-top effects sequences really stand out. For example, in contrast to images of a 600-foot-tall toddler, his mother appears on camera remarking with a straight face that it became a problem when Luca started growing out of his clothes. The effect is similar to a comedian who appears onstage with a straightman—each takes the other to a higher level of potency, and as a result, each punch line delivers a stronger payoff. (Think about the classic comedy teams of Abbot and Costello or Martin and Lewis.)

Placing an interview clip after your opening title in Final Cut Pro

To insert an interview clip in Final Cut Pro, do the following:

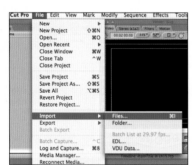

1. Drag the interview clips (labeled *interview1.mov*, *interview2.mov*, *interview3.mov*, and *interview4.mov*) from the StartToFinish DVD to your computer.

2. Select File→Import→Files…. The Choose a File window opens.

3. Navigate to the interview video clips you just dragged into your computer.

4. Click *interview1.mov*, which is the first in the list. Final Cut Pro selects the file.

5. Shift-click *interview4.mov*, which is the last file in the list. Final Cut Pro selects the file, along with the other two files, which appear in between.

6. Click Choose. Final Cut Pro imports the files and closes the Choose a File window. The clips now appear in the Browser window.

7. Double-click *interview1.mov* in the Browser window. The clip opens in the Viewer.

8. Play the clip in the Viewer. The clip opens with Luca singing (we hear him over a blank screen) and then we hear his mother in an interview (we first see a still image of a very young Luca, and then cut to his mom talking on screen).

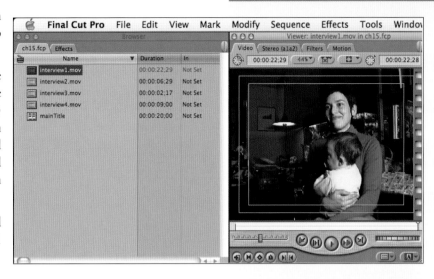

9. Move the Playhead to the end of the title graphic on tracks V1 and V2.

10. Click the Insert Edit button in the Canvas. Final Cut Pro adds the interview clip to the Timeline immediately after the title graphics.

11. Drag the Playhead through the sequence to preview what you've created so far, or render the composite. The opening zoom segues nicely into the title text, which is followed shortly by the first interview clip (this clip is also the first part of the body of the film). Because the interview clip begins with Luca singing over a black screen, it creates a natural bridge between the title and the following sequence.

12. Save your work.

NOTE

Because the sequence is now considerably longer, you may not be able to see all the clips at once in the Timeline. To zoom out to see more of the Timeline, hold down the Command key and press the – key. Each time you press the – key Final Cut Pro zooms out to a wider view of the Timeline. To zoom back in, hold down Command and press the + key.

Placing an interview clip after your opening title in After Effects

To insert an interview clip in After Effects, do the following:

1. Drag the interview clips (labeled *interview1.mov*, *interview2.mov*, *interview3.mov*, and *interview4.mov*) from the StartToFinish DVD to your computer.

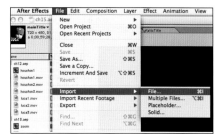

2. Select File→Import→File.... The Import File window opens.

3. Click *interview1.mov*, the first interview clip in the window, to select it.

4. Shift-click *interview4.mov*, the last clip in the window, to select it along with the other clips in between.

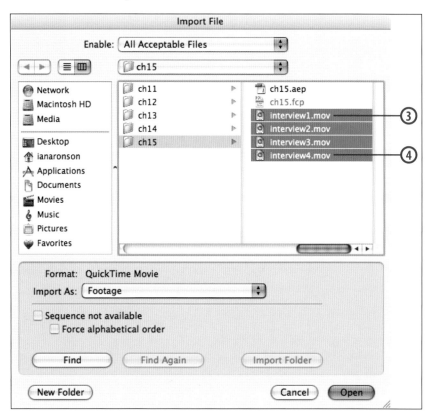

5. Click Open. After Effects imports the clips into your project and closes the Import File window. The clips now appear in the Project Window.

6. Double-click the *interview1.mov* clip in the Project window. The clip opens in a new window.

7. Click the Play icon in the *interview1.mov* window. The clip opens with Luca singing (we hear him over a blank screen) and then we hear his mother in an interview (we first see a still image of a very young Luca, and then cut to his mom talking onscreen).

8. Close the *interview1.mov* window by clicking in the upper-left corner.

9. Drag the *interview1.mov* clip from the Project window into the Timeline, to the right of the existing clips. After Effects adds the clip to the composition.

10. Move the Current Time Indicator through the Timeline to preview your new composite, or simply render the composition. After the opening zoom fades to black, the title text scales its way onto the screen, and the edited still image of Luca's face pops onto the screen. The interview clip, which is also the first part of the main body of your film, immediately follows.

You've crafted a very slick opening sequence and linked it organically to the start of your story. Very nice job.

11. Save your project.

Earlier in this book, you created a series of intricately detailed effects and rendered them to create high-quality digital video files. In this chapter, you learned how to weave them together using custom tailored segue transitions specifically suited for the look and feel of each composite. The exercises in this chapter (as well as in the rest of the book so far) have walked you through a series of very sophisticated, professional-quality techniques, and I hope you feel a sense of accomplishment.

If you'd like to further refine your skills, use the additional footage in the *Ch12* folder to create more effects sequences. There's no better way to improve your abilities than through practice. Once you've got some more effects that you're happy with, use the techniques you learned in this chapter to bring the new effects sequences into your project—you can even combine them with the additional interview clips in the *Ch15* folder to make an entire film, complete with a beginning, middle, and end. (Don't forget to add credits; as you learned in Chapter 14, you can make use of scrolling and crawling titles.)

The next chapter examines color correction, which is the process of ensuring the colors in your film look exactly the way you want them to. You now have the ability to make any film you want—the remainder of this book shows you how to polish your work and bring it to an even higher level of quality.

Color Correction

16

Color is one of the first things people notice when they watch your film, whether they're conscious of it or not. As explained in detail later in this chapter, color functions on both aesthetic and emotional levels, deeply and immediately influencing the way each viewer responds to your work. Color is also a particularly complex technical process, and like many other aspects of filmmaking, if your project contains problems with color, your audience will notice right away.

The human eye and brain constantly work together to compensate for color differences in the world around us. If you leave a room lit by a standard household light bulb and walk outside into midday sunshine, you might not notice that the color cast of the ambient light around you changes from a yellow tint to a blue tint, but a video camera would certainly record the difference. (For more on color temperature and the importance of white balance, see Chapter 4.)

Color varies depending on lighting conditions, camera equipment, and even time of day. Audiences notice when the color cast of a shot doesn't match the color in the shots that come before and after. Skilled directors can use color differences to separate multiple threads in a story—as Zhang Yimou did in *Hero* and Steven Soderbergh did in *Traffic* (both films are described in detail later in this chapter)—or to accentuate an emotional motif. Most of the time, however, audiences read differences in color cast as a technical mistake and lower their opinion of a film accordingly, even if they aren't consciously aware of what sparked their negative reaction.

Cinematographers can compensate for color differences during production by making careful white balance adjustments and, if possible, checking their work on an external NTSC field monitor. During postproduction, editors can use color-correction tools to match the color from one shot to another, or to create a desired effect.

In addition to aesthetic considerations, an NTSC or PAL video signal can only display a limited range of color information. If the color in your project falls outside an acceptable range—for example, if a color is too bright or too

In this chapter

How Color Functions in Video, an Overview

Audiences' Subconscious Response to Color, and How You Can Make Use of It

Making Your Video Broadcast Safe

intense—it won't display the way you want and can even make an image unrecognizable or interfere with the audio tracks in your work.

How Color Functions in Video, an Overview

In digital editing applications, color information is measured in two ways luminance (or *luma*), which refers to brightness, and chrominance (or *chroma*), which refers to what most people think of as color—for example, red or blue. By making adjustments to the luminance and chrominance levels in a video clip, editors can ensure that color information will reproduce as intended on a variety of video monitors and in a television broadcast. (This is analogous to "mastering" a record in the music world.) If you've ever watched a well-produced, big-budget television program where images look really rich and full, and then watched a show on public access where the images look really thin in comparison, the difference is probably due, at least in part, to color correction.

Television networks spend lots of time and money to ensure color levels fall into an acceptable broadcast-safe range. Broadcast-safe levels are also referred to as legal levels and correcting non-broadcast-safe color is sometimes referred to as making the levels legal. There's no actual law or statute, the term simply describes levels that are acceptable for broadcast. Even if you never expect your work to find a broadcast audience, broadcast-safe color levels are still important, because they ensure your colors will reproduce the way you want and without causing technical difficulties for your viewers.

Luminance

Luminance measures how bright the information in a video signal is, on a scale with the darkest possible black at one end and the brightest possible white at the other. (Final Cut Pro uses a scale measuring from 0 to 100; After Effects uses a scale measuring from 0 to 120. The darkest black measures 0 in both applications.) Any brightness values at the outside ranges of the scale are not considered broadcast safe, meaning they will distort on a television monitor. As described in Chapter 11, the RGB color space used by image-editing programs such as Adobe Photoshop tolerates a wider range of color than an NTSC video signal. As a result, bright whites created in Photoshop or another RGB program may not be broadcast safe. Non-broadcast-safe whites are called superwhite, and can cause distortion in images and even bleed onto the audio track, creating an annoying hum or a buzz. Likewise, RGB images can contain blacks that are too dark to display in video. When black levels in a video image are too dark, details get lost—for example, shadows and backgrounds blend together and contours disappear.

SIDEBAR

RGB Design and the NTSC Signal

RGB color values extend beyond the range of what the NTSC video signal can safely reproduce. As a result, the darkest possible black you can create with a program like Photoshop or Illustrator is too dark to reliably display on an NTSC monitor. Similarly, the brightest possible white you can create with Photoshop or Illustrator is too bright.

The solution is to create images that don't use the darkest or the brightest possible RGB values.

The image at right shows the color palette in Photoshop, adjusted to create a broadcast-safe black. The red, green, and blue channels are each set to 16, which will produce the darkest possible broadcast-safe black. (Photoshop enables a designer to create a darker black than the one pictured here, but it would not be broadcast safe.)

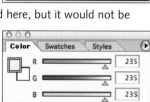

This image shows the Photoshop color palette set to create a broadcast-safe white. The red, green, and blue channels are each set to 235, which will produce the brightest possible broadcast-safe white. Photoshop allows a designer to create brighter whites, but they would fall outside the broadcast-safe range.

For more information on creating broadcast-safe RGB images, and the additive color process that computer monitors and video monitors both use to reproduce color, see "Creating images in Photoshop for use in digital video" in Chapter 11.

Chrominance

Chrominance, which measures the color information in a video clip, breaks down into two parts: hue and saturation.

Hue

Hue defines a color in terms of its location on a color wheel. For example red, green, and blue, the primary colors in a video signal, are located at –13 degrees, –151 degrees, and 103 degrees, respectively. Locations on the wheel are measured in degrees because of the wheel's circular shape. See Figure 16-1.

Using hue to give each color a numeric value provides an accurate way to identify specific colors. I often begin class discussions about color by asking students to think about the color red. I then explain that although everyone in the room might indeed be picturing a red object, each person may actually be thinking of a slightly different color. One might picture the red

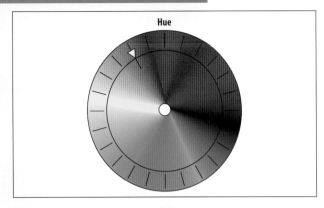

Figure 16-1. The hue of a color measures its location on a color wheel. The color wheel can be used to provide a specific numeric value, or hue, for any color in a video signal.

in a Coca-Cola logo, another might picture the red of a brick building, others might picture the red of an apple or tomato (which can each vary greatly). Identifying a particular color with a numeric value ensures that when editors discuss a specific hue during the color-correction process, everyone shares the same understanding.

Saturation

Remember the term "saturation" from junior high school science class? When something is saturated it holds all it can. The concept of saturation works more or less the same way in video: it describes how intense a color is —think of a highly saturated red as holding all the red color information it possibly could.

Saturation uses a different color wheel to measure the location of a color in terms of its distance from the center (Figure 16-2). A blue color with a saturation level closer to the edge of the circle would be a dark blue. A blue of the same hue, located closer to the center of the circle, would be a pale blue, or a light blue; this blue is referred to as desaturated. Highly saturated colors are rich in color information; they're often described as *vibrant*.

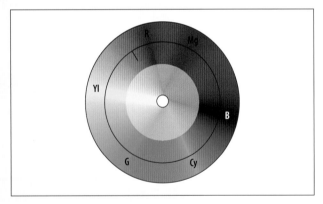

Figure 16-2. In the saturation color wheel, colors are located in the same places around the edge as in the wheel measuring hue. The difference is, highly saturated colors closer to the outside of the saturation wheel contain appear richer in color than those nearer to the center.

Just as there are broadcast-safe standards for luminance, there are standards for saturation. Computer monitors can handle highly saturated colors far beyond the range of what an NTSC monitor or television can safely reproduce. If saturation levels fall outside an acceptable range, colors can bleed into one another, or in particularly bad situations, create audio problems.

Broadcast outlets have very specific standards of what they consider acceptable luminance and chrominance values. If a program doesn't meet their broadcast standards, television organizations will often return the master tape to the producer and ask him to fix the problems before airing the work. PBS has particularly strict broadcast standards, which sometimes include financial penalties for producers who submit programming that does not pass muster during technical evaluation. Final Cut Pro and After Effects both offer tools, discussed later in this chapter, to ensure your work is broadcast safe.

Audiences' Subconscious Response to Color, and How You Can Make Use of It

Humans have a complex and visceral understanding of color, and it impacts the way we respond to just about everything in the world. Color shapes and informs not only our aesthetic sensibilities about visual objects but also helps determine our enjoyment of important everyday activities such as eating—can you imagine sitting down to a pleasurable meal of gray food?

Similar influences are at work when people watch a film. People have an instinctive understanding of what colors mean, so the colors in your film can spark specific emotional reactions in your viewer. For example, an image of a pale strawberry would not look nearly as appealing to your audience as a bright red strawberry with the same hue but a higher saturation value. Depending on the tone of your project you might want the audience to find food unappealing, for example in an apocalyptic horror movie, but for a cooking show, more saturated colors will bring more positive results.

Audiences are especially well attuned to color values when it comes to flesh tones, and particularly so to the tones in people's faces. The color in a person's face provides an instant indication of their health: rosy cheeks indicate physical well being, while green implies seasick—or worse. (This holds true for people of different ethnic backgrounds and skin tones. A person with very dark skin can appear noticeably flushed, pale and peaked, or green and queasy.) A filmmaker can manipulate color in postproduction to make a person appear more or less likable on screen. Viewers will respond well to a child with a "warm" face and hints of red in her flesh tones but find themselves wary of a person whose skin appears to have a greenish tint.

Adding a color cast to a scene

Directors often add a color cast to a sequence in order to emphasize a particular emotional value. Love scenes are often tinted red, a color symbolizing romance and affection in a number of cultures, while blizzard scenes and stark police dramas are often tinted a pale shade of blue to indicate a distinct lack of physical or emotional warmth. (In the film *In the Mood for Love*, director Wong Kar Wai skillfully uses several different hues and saturations of red to explore themes of love and sentimentality.)

The film *Amélie* uses a particularly yellow tint to cast a hopeful and sunny glow on what might otherwise be seen as the main character's lonely and dreary life. Even in shots when it's raining, everything on screen appears with a warm yellow tint. Interestingly, *Amélie*'s director Jean-Pierre Jeunet uses a similar yellow tint in his next film, *A Very Long Engagement*, to alternately explore visual themes of sentimentality and the misery of World War I trench warfare. Two of Jeunet's earlier works, *City of Lost Children* and *Delicatessen*, are decidedly darker, both in terms of color palette and subject matter. (I won't ruin the surprise by describing it for you, but *Delicatessen* has what may be my favorite opening sequence of all time.)

Filmmakers also add distinct color casts to different sections of a film to help audiences keep track of different strands in a story. In *Traffic*, director Steven Soderbergh used a yellow color cast for the plot line that follows

Using Color to Create a Positive Emotional State

In *Emotional Design*, Donald A. Norman, a professor of computer science and psychology at Northwestern University, writes that "bright, saturated hues" and "comfortably lit places" put people in a positive emotional state in which they're more receptive to having a positive emotional experience. According to Norman's book, color and lighting function on a subconscious level, shaping people's opinions before they've had a chance to make a conscious evaluation.

Norman also lists conditions that produce "automatic" negative states, such as "darkness" or "extremely bright lights or loud sounds." Depending on the emotions you'd like to spark in your viewer, you can color correct your footage accordingly, so that viewers are predisposed to react in a particular way.

Norman writes that humans are "automatically programmed" to respond well to certain environmental factors and to react negatively to others. Keeping an audience's subconscious responses to color in mind as you color correct your film can help you make a very powerful project.

Benicio Del Toro as an underdog Mexican law enforcement officer, and a blue color cast for the subplot following Michael Douglas's attempts to grapple with his teenage daughter's drug problem. Memento juxtaposes color footage with black and white. Parts of the enfolding story are shown in color, while the director withholds crucial information from the audience and slowly reveals it in a series of black and white sequences. As digital technology provides filmmakers with increasingly powerful color-correction tools, directors are adding even more color effects to their work. *Hero*, a martial arts action film by Chinese director Zhang Yimou, contains five subplots, each with its own color cast. The film tells the same story from the point of view of three different characters, and each story gets its own color: red, blue, or white. The director then adds flashbacks with a green tint, as well as a fifth series of "true" sequences filmed in shades of black (for example, the members of the emperor's court wear black armor that perfectly matches the immaculate palace decor).

The film is in many ways similar to Akira Kurosawa's 1950 classic *Roshomon*, which also tells one story from three perspectives. Shot in black and white, *Roshomon* provides a seminal meditation on themes of truth, experience, and perception. *Hero* explores similar areas, using sophisticated techniques to color each storyline both in terms of its plot and in the appearance of its visual elements.

"Every story is colored by personal perception," the film's cinematographer Christopher Doyle told the *New York Times* (Doyle also shot *In the Mood for Love*). The filmmakers chose green for the flashbacks "because we ran out of colors." According to Doyle, they had already used red, blue, and white in other storylines, so green was all they had left, since "You're not going to do anything in orange or pink." (As the director of your film, you may decide you like orange and pink, in which case you can color your work accordingly—as I've mentioned throughout the book, the director of a film is the captain of the ship.)

For more information on connecting an external NTSC monitor to your system, see Chapter 10. The following exercises walk you through the process of adding a color cast to your video using Final Cut Pro or After Effects.

SIDEBAR

The Importance of Calibrating a Video Monitor

Before you start the color-correction process, it's important to calibrate your video monitor to display a standard set of colors. The colors displayed by different monitors can vary greatly, so without calibration, there's no guarantee that what looks good on one monitor will look anywhere near as good on another.

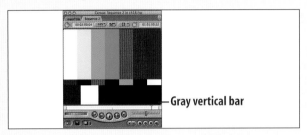
— Gray vertical bar

To calibrate a video monitor, send color bars to your video monitor. Some professional monitors generate their own color bars; if your monitor doesn't, you can send color bars to your monitor from your camera or your editing system.

To send color bars from your camera:

- If your camera generates color bars, run color bars in your camera. (Not all DV cameras can generate color bars. Some higher-end prosumer camcorders and many professional models do, and the process of generating color bars varies widely by model. Check your camera's operating manual for more details.)

- Connect the video output of your camera to the video input of your monitor.

To send color bars from Final Cut Pro:

1. Select the Effects tab in the Browser window.

2. Click the twirl-down icon next to Video Generators.

3. Double-click Bars and Tone (NTSC). Final Cut Pro opens a two-minute clip of color bars in the Viewer window.

4. Edit the color bars clip into a new Sequence. If you have an external video monitor connected to your Final Cut Pro system, the color bars will display on your monitor.

5. Near the bottom right corner of the color bar display in your monitor, you should see a light gray vertical bar. Adjust the brightness control of your monitor so that the gray bar is just barely visible.

6. Adjust the contrast control of your monitor so that the bar is fully visible, and then slowly re-adjust the contrast to make the bar darker. Continue to adjust the contrast setting until the bar becomes just barely visible. When the bar displays in your monitor as dark gray and just barely visible against the black areas surrounding it, your monitor is properly calibrated.

Using Final Cut Pro to change the color of your video

This exercise uses the Color Corrector filter to add a color cast to the *interview01.mov* clip.

1. Open the mainTitle sequence in the Timeline.

2. Drag the Playhead to 01:00:30;10 in the Timeline (or you can enter 01:00:30;10 in the timecode field in the top left of the Timeline window). The head of *interview01.mov* clip contains a still of a very young Luca, and the rest of the clip shows his mother in an interview—frame 01:00:30;10 is the first time she appears onscreen.

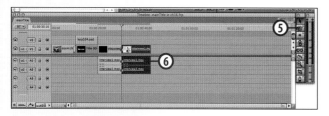

3. Click the Razor Blade tool in the Tool palette. The Razor Blade tool enables an editor to separate a clip in the Timeline into two separate clips. In this example, separating the *interview01.mov* clip permits you to make changes to the interview footage of Luca's mom without disturbing the color balance of the still image that appears at the head of the clip.

4. Using the Razor Blade tool, click the *interview01. mov* clip at the Playhead's current location. Final Cut Pro splits the clip in two.

5. Click the Selection tool in the Tool palette.

6. Click the second part of the *interview01.mov* clip in the Timeline to select it.

7. Select Effects→Video Filters→Color Correction→Color Corrector. Final Cut Pro applies the filter to the portion of the clip in which Luca's mom appears on screen.

8. Double-click the clip in the Timeline. Final Cut Pro opens the clip in the Viewer.

9. Click the Color Corrector tab in the Viewer to access the filter's controls.

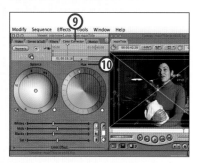

10. Drag the outer ring of the Hue control to the right to add a red tint to the clip. A slight red tint adds a rosy glow to the face of Luca's mom (who, by the way, is named Deirdre). Because computer monitors display color differently than video monitors, viewing the results of your work on a dedicated NTSC monitor becomes particularly important when you're making color adjustments. For more information on the benefits of using an NTSC monitor and how you can connect one to your editing system, see Chapter 10.

- If you drag the outer ring of the Hue control a few degrees to the left of the default value, Final Cut Pro adds a yellow tint to the shot.

- To reset the Hue value to its default setting, click the reset button to the immediate right of the Hue control.

11. Save your work.

Using After Effects to change the color of your video

This exercise uses the Hue/Saturation effect to add a color cast to the *interview01.mov* clip.

1. Drag the Current Time Indicator to 01:00:30;10 in the Timeline (or you can enter 01:00:30;10 in the timecode field in the top left of the

---

NOTE

You can make color footage black and white by completely desaturating the clip. To do this, drag the Saturation slider (at the bottom of the Color Corrector window) all the way to the left.

Timeline window). The head of *interview01.mov* clip contains a still of a very young Luca, and the rest of the clip shows his mother in an interview—frame 01:00:30;10 is the first time she appears onscreen.

2. Click the *interview01.mov* clip in the Timeline to select it.

3. Select Edit→Split Layer. This command separates the interview clip you selected into two parts, and places each on its own layer.

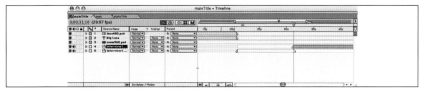

In this example, splitting the *interview01.mov* clip onto two separate layers permits you to make changes to the interview footage of Luca's mom without disturbing the color balance of the still image that appears at the head of the clip.

4. Click the part of the clip that contains the interview with Luca's mom, and then select Effect→Adjust→Hue/Saturation.

After Effects adds a Hue/Saturation effect to the selected layer and opens the Effect Controls window. (If the Effect Controls window doesn't open, or if you accidentally close it, you can access the window by selecting Effects→ Effect Controls.)

5. Drag the Master Hue dial a few degrees to the left to add a slight red tint to the clip. Luca's mom (Deirdre), now appears with skin tones that make her look subtly more healthy than she did before the effect.

• You can also adjust the Master Hue setting by entering a value directly into the numeric field above the dial. Entering a positive number moves the dial to the right of the default value, entering a negative number moves the dial to the left of the default. Alternately you can scrub the degree (°) number in the Master Hue to the correct setting.

• Moving the dial just to the right of the default would give the clip a green tint.

Because computer monitors display color differently than video monitors, viewing the results of your work on a dedicated NTSC monitor becomes particularly important when you're making color adjustments. For more information on the benefits of using an NTSC monitor and how you can connect one to your editing system, see Chapter 10.

6. Save your work.

Color Matching and Darkness in *Open Water*

The husband-and-wife filmmaking team Chris Kentis and Laura Lau used a Sony VX2000 and a PD-150 to shoot the very scary *Open Water,* which was transferred to film and ultimately released on more than 2,000 screens across the country. *Open Water* brings audiences the fictional story of two scuba divers who are inadvertently left at sea after their tour boat returns to port without them. Kentis and Lau told *Filmmaker* magazine they shot on DV because the low cost and flexibility enabled them to complete the project entirely on their own—without the help of producers, financial backers, or even a crew. "We wanted total control," Lau said. By shooting prosumer digital video, Kentis and Lau also exploited a strength of the medium: the immediacy of DV makes audiences feel like witnesses to actual events unfolding on screen.

The realism of the film is heightened by the fact that the two lead actors really are in the open ocean, surrounded by sharks. For much of the film, sharks swim uncomfortably close to the actors—and it's real. The small scale of production resulted in a great experience for the audience, and even for the actors. Blanchard Ryan, who plays the female lead, told the *New York Times*, "It was the most magical working environment—two actors, two filmmakers, no hair and makeup . . . just get in the water and do the work."

The approach also created technical challenges. The actors were in the ocean with real sharks, unprotected by any type of cage. To reduce the risk of an actual attack the actors kept very still as the sharks swam around them feeding on chunks of tuna that had been thrown into the water. The filmmakers then shot the dialog scenes separately, away from the sharks, along with action shots in which the actors kicked or moved suddenly in ways that might have been too dangerous to try with sharks feeding around them. When they returned to New York after they finished shooting, the filmmakers brought their footage to a specialist who used After Effects to color correct their material and match the lighting conditions from shot to shot. After a significant amount of work, they made the footage look like it was recorded in one uninterrupted series of events all shot in the same location at the same time of day. (For details on why it can be difficult to match the color and lighting of shots recorded at different times of day, see "Continuity and Detail Management" in Chapter 3.)

The filmmakers also do a good job using color early in the movie, to make an otherwise romantic sunset look deeply ominous. Shortly after the lead characters begin their vacation, and before they start their ill-fated dive, the audience sees a sequence of beauty shots establishing the look and feel of the Caribbean resort town the characters are visiting. One of the shots depicts a sunset in which the light reflecting off the water turns the sea blood red. Through the effective use of color, the filmmakers have turned a benign postcard shot into an agent of foreboding. An orange-colored sunset connotes feelings of paradise—this sunset clearly conjures images of blood.

Kentis and Lau also use darkness to great effect toward the end of the film. After a day of being stranded in the open ocean, the protagonists find themselves still stranded in the middle of the night in complete darkness. Rather than providing an artificial light source, since there would be none on a cloudy night at sea, the filmmakers leave the screen black for disconcertingly long periods of time, illuminating the action only with flashes of lightning. Donald Norman writes that darkness can trigger a negative emotional state (you may have figured that out on your own—kids love nightlights for that exact reason). In *Open Water*, the filmmakers use darkness very well, creating some especially scary sequences.

Making Your Video Broadcast Safe

For years, the thought of having to make something broadcast safe struck fear into many a video editor. No one argued that making video levels broadcast safe was unimportant, or that it wasn't worthwhile, but checking levels and making them legal required specialized equipment and a specially trained operator. Even with the introduction of the Avid digital editing system in the mid 1990s and its complete array of finishing capabilities, many editors still outsourced color correction to experts at a postproduction facility.

Fortunately, both After Effects and Final Cut Pro offer effects to bring both the luminance and chrominance levels within broadcast-safe ranges. This section shows you how to use both. As mentioned earlier in this chapter, ensuring your work is broadcast safe helps avoid technical problems even if you never plan to broadcast your project.

Using the Broadcast-safe filter in Final Cut Pro

To apply the Broadcast-safe filter:

1. Select the clip you'd like to make broadcast safe (you can only apply the filter to one selected clip at a time).

2. Select Effects→Video Filters→Color Correction→Broadcast Safe. Final Cut Pro applies the filter to your clip and "clamps down" any unsafe levels to bring them within an acceptable range. Unless your uncorrected footage contained images that were significantly too bright or especially over saturated, you probably won't see a noticeable change in your composite after Final Cut makes the clip broadcast safe.

3. Save your work.

How to Determine Whether Footage Is Broadcast-Safe

To check the brightness and color levels in a video clip, technicians use a waveform monitor and a vectorscope. A waveform monitor measures brightness, or luminance, and a vectorscope measures color, or chrominance.

Waveform monitors and vectorscopes can be either hardware- or software-based. A hardware-based waveform monitor or vectorscope is a standalone piece of equipment connected to an editing system. Some video editing and compositing applications, such as Final Cut Pro, ship with a software-based waveform monitor and vectorscope built into the program. After Effects does not ship with a waveform monitor or vectorscope, but you can purchase a plug-in that expands After Effect's capabilities to include them. (For a full list of third-party plug-ins available for After Effects, visit *http://www.adobe.com/products/plugins/aftereffects/main.html*.)

A well-equipped postproduction facility will have both hardware- and software-based waveform monitors and vectorscopes available, as well as experienced staff members who know how to use them. As mentioned in the sidebar "Why You Might Want to Hire a Professional Colorist" later in this chapter, working with a professional to color correct your footage and make sure it's broadcast safe can help raise the quality of your film to a highly professional level. This may become particularly important if you're seeking theatrical or broadcast distribution (although some filmmakers negotiate deals in which a distributor pays for color correction after acquiring a film).

Broadcast Safe, Rendering, and Nesting

After applying the Broadcast-safe filter in Final Cut Pro or the Broadcast Colors effect in After Effects you'll need to render your sequence before you can see the changes in the playback. See Chapter 11 for more details on rendering.

In both applications, the color correction you apply changes only the individual clip you have selected. To simultaneously apply a color correction filter or effect to all the clips in a composite:

- Nest the entire composite in a new Sequence (Final Cut Pro) or in a new Composition (After Effects).

- Add a color-correction filter or effect to the nested composite. The filter or effect will simultaneously apply to all the clips in your composite.

For more information on nesting a complete Sequence or Composition within another, see the sidebars "Nesting vs. cut and paste in FCP" and "Nesting a Composition in After Effects" in Chapter 15.

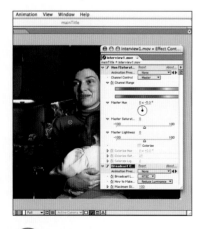

Using the Broadcast Colors effect in After Effects

To apply the Broadcast Colors effect:

1. Select the clip you want to adjust by applying the Broadcast Colors effect.

2. Select Effect→Video→Broadcast Colors. After Effects applies the effect and opens the Effect Controls window. (If the Effect Controls window doesn't open, or if you accidentally close it, you can access the window by selecting Effects→Effect Controls.)

The default setting for the Broadcast Colors effect lowers the highest luminance values in your clip to 110, which the After Effects documentation describes as "conservative." This means the setting is designed to make your work legal without significantly changing its appearance.

If you want to adjust the luminance levels to bring them further inside the broadcast-safe range you can enter 100 (or even 90) in the Maximum Signal field at the bottom of the Effect Controls window.

To adjust the saturation values in the same clip, you can add another instance of the Broadcast Colors effect to the same clip.

3. Select Effect→Video→Broadcast Colors. After Effects applies a second instance of the effect to your selected clip.

4. Click the How To Make Color Safe pull-down menu and select Reduce Saturation. By default, After Effects adjusts the highest saturation levels in your clip to 110. You can lower the value, to bring the saturation levels further inside the broadcast-safe range, by entering a lower number in the Maximum Signal field at the bottom of the Effect Controls window. (The example shows the effect controls set to lower the maximum saturation value to 100.)

5. Save your work.

Why You Might Want to Hire a Professional Colorist

According to Filmmaker magazine, when it was time to color correct *Open Water*, Kentis and Lau took their footage to a professional facility. Although Kentis, who earned a living editing trailers for feature films, cut *Open Water* together himself using Final Cut Pro, he apparently decided there are times when professional services are worth the expense.

Professionals offer an ease and skill that enables them to work very quickly and get excellent results. In addition, they have access to professional equipment beyond the prosumer price range. Even if a professional charges a high hourly rate, he might be able to finish a job quickly, saving you considerable money in the long run—not to mention aggravation.

I used to work with a man who was set on renovating his home himself, even when he didn't know what he was doing. He regularly told me stories about opening up walls to attempt plumbing or light electrical repairs only to create a mess he couldn't clean up. He would then call

a handyman to come to his home and bail him out. For a fee, the handyman would happily oblige, after shaking his head at the fiasco my colleague had created.

Unfortunately, plenty of filmmakers operate the same way. Hiring a professional before you get yourself into technical problems is much cheaper than paying someone to undo a mistake you can't fix.

To some people, the actual process of completing a media project is just as important as the end result. For example, I have a good friend who always enjoys doing his own technical work, even if it takes longer and costs slightly more than outsourcing part of a project. He always does a very good job and derives great satisfaction from finishing things on his own.

As the director of your project, you have to choose the path that's right for you and the methods that best meet the needs of your film.

This chapter explored color from aesthetic, psychological, and technical perspectives. It also walked you through the process of intentionally adding a color tint to a clip and then adjusting the luminance and saturation levels of your footage to ensure everything not only remains broadcast safe, but looks exactly the way you want.

Attentively refining the color in your film enables you to take your work far beyond the bare minimum of technical competence—in short, a good use of color can make the difference between a merely acceptable project and something that's truly outstanding.

Creating the greatest film possible means focusing on each detail and reworking all the elements in your project until you've made everything as powerful as you can. Up to now, you've concentrated on the images in your work; the final two chapters of this book are devoted to maximizing the effectiveness of your audio. Chapter 17 explores the importance and intricacies of creating a rich sound design, a multilayered audio montage that delivers just as much of an impact as the visuals in your project. Chapter 18 explains how to add music to a project (without violating copyright laws) and also examines ways to fine-tune all the layers of audio in your film.

You're almost ready for your red carpet premiere.

The Concept of Sound Design

17

The audio in a film doesn't get there by itself. Just as well-edited images and carefully executed composite sequences don't instantly materialize in a project, neither does the audio that accompanies them. Some beginning filmmakers treat audio as an afterthought; others ignore it almost completely. Film and video are, after all, *visual* media, so some filmmakers wonder why they even need to think about audio.

Accomplished filmmakers, however, know that an audience's experience watching a film depends largely on the success of the film's sound design. Audio, or its conspicuous absence, can shape or even determine a viewer's reaction to what he sees on screen. Audiences can literally feel an explosion when a sound designer adds a thundering low-frequency rumble to the film's audio track. Viewers are similarly drawn into the physical environment of a sequence when a sound designer adds layers of ambient sound that might include traffic noise, the background hum of overheard conversation, the rhythmic mechanical clangs of an industrial environment, or an infinite number of other elements depending on the nature of the production.

The sound design of a film refers to the way audio has been recorded, edited, layered, and mixed to provide the audience with the richest possible experience. A good sound designer works closely with the director and editor to determine exactly which sounds to emphasize in a sequence, what to replace, what to omit, and what to manufacture in order to create a film that not only sounds as good as real life, but sounds like a good movie. The world never looks as good in everyday life as it does in a well-crafted film, and it doesn't sound as good either. That's why good cinematographers and audio engineers can always make a living, and it's also why we pay good money to go to the movies.

There is perhaps no filmmaker more adept at developing a film's rich aural texture than Walter Murch. Working for years with director Francis Ford Coppola, Murch earned great recognition as the sound designer for *Apocalypse Now* (which he also co-edited) and for *The Conversation*, which may be the most audio-driven film of all time. He also did the sound for George Lucas' *American Graffiti*, in which he made an innovative use of period music blended into background sound (when characters in the movie listened to Wolfman Jack on their car radios, it actually sounded to the audience like the sound was coming from a car radio). Murch later went on to edit the multi-award–winning film *The English Patient* for which he received Academy Awards in both editing and sound. Following the success of *The English Patient*, Murch released a version of *Touch of Evil* in 1998 that he restored to the original specifications of director Orson Welles. (After Welles delivered his final version, executives at Universal made a series of changes—including hiring a new director to add scenes—that Welles felt destroyed the heart of his film.) Murch's restored version follows a 58-page memo that Welles sent Universal detailing his objections to the studio's changes. Murch's 1998 *Touch of Evil* retains the legendary opening shot, mentioned in the first chapter in this book, and augments it with a sound design unrivaled in detail, complexity, and understated technical sophistication. Murch has in many ways set the standard for high-quality sound design, and this chapter examines a number of his films to demonstrate various sound design principles and techniques.

SIDEBAR

The Juxtaposition of Audio and Picture

The first three minutes of *Apocalypse Now Redux,* the reissued director's cut of the cinematic classic, employs a dreamlike juxtaposition of audio and images. The film opens with a black screen and the slowed down, echoing sound of a helicopter. The screen fades up to a static shot of trees in the distance, and a helicopter passes through the frame in close-up, followed by the highly saturated yellow smoke of a signal flare that floats gently across the screen. After a few seconds, the highly styled helicopter sounds dissolve into the opening bars of "The End" by the Doors. As Jim Morrison begins his vocals, the trees erupt into a fiery explosion, but there's no explosion or warfare sounds, in fact, no sound at all other than the slow, ironically peaceful music.

The combat visuals gradually dissolve into shots of the main character, played by Martin Sheen, passed out in a hotel room with a gun on the pillow next to him. As the combat scene begins to fade away, the sound of helicopter rotor blades returns, but instead of accompanying a shot containing a helicopter, Coppola pairs the sound with an image of the ceiling fan spinning in Sheen's hotel room. As Sheen's character awakens, the up close and almost tangible sound of the helicopter dissolves into the sounds of another helicopter, this one echoing into Sheen's hotel room from somewhere outside.

This use of sound deliberately throws the audience off balance. Sounds the viewer expects to hear, such as helicopters during the combat sequence, are noticeably absent, and show up inexplicably in the next shot of Sheen's hotel room. This intentionally disorienting sound design brings the audience into the troubled mind state of Sheen's character and sets the overall confusing and almost hallucinatory tone of the film.

Layering Audio Tracks

Sounds don't occur in isolation. The audio that people hear in life and in film contains multiple sounds layered together to form multitextured aural compositions. Even in a quiet room, there's sound. When I teach a course that involves sound recording and editing, one of the first things I do is ask everyone in the room to be quiet and think about the sounds they hear. Just as the human brain filters out differences in color (see Chapters 4 and 16), the human brain also does an impressive job filtering out background sound; the brain is, in fact, so efficient at this that most people never even notice background noise.

After a brief pause to let students identify the background sounds in a room, I then ask everyone to list the sounds they heard when no one was speaking. Common sounds include the hum of a computer (both hard drives and monitors create sound), electric lighting (fluorescent lights often buzz), any type of heating or cooling system, chairs moving on a hard floor surface, even footsteps outside or on the floor above. People spend their whole lives tuning out these sounds in everyday life, but if any one were missing in a film, audiences would notice immediately. At the same time, if any one of these elements appeared in a film at a level loud enough to overshadow the others, audiences would notice that, too, and they'd think it was a really annoying mistake.

Adding layers of audio in the Timeline

Layering audio in the Timeline is similar to layering visual elements in a composite sequence. In the sequence in Figure 17-1, audio elements are placed on different tracks so they can be adjusted independently. Dialog elements are placed on tracks A1 and A2. Room tone, or the ambient sound of the room, is placed on A3 and A4. Specific sounds that have been recorded separately and are being added to augment part of the sound design are added to tracks A5 and A6. Because the recordings in this illustration are all stereo audio clips, each one takes up two tracks.

The layering of the clips in this example follows an accepted pattern. Dialog clips are generally placed in the first two tracks, while music and effects are placed on additional tracks known, appropriately, as the M and E tracks (M and E stands for music and effects). This arrangement becomes especially helpful if you decide to create a foreign language version of a project: if all the dialog and voice over are on the first two tracks, you can simply replace the material on tracks A1 and A2 and keep the rest of your sound design intact.

Figure 17-1. Layering audio in the Timeline.

Because the world contains multiple layers of sound audible to a listener at any time, creating a good soundtrack for a film requires using multiple recordings and layering them together. Imagine you're in a classroom, and as a teacher explains a concept to the class, she writes on the board. Think about the sounds in the room at that time: in addition to ambient sound (also known as *room tone*, which includes things like the buzz of lighting fixtures and the sound of steam heat coming through a radiator), you'd hear the sound of the teacher's voice and, of course, the sound of the chalk making contact with the blackboard.

The toughest challenge arises when capturing the sound, because a good recording of any one sound element probably won't yield a good recording of the other ambient sounds. For example, a good recording of the teacher's voice would position the pick-up pattern of a microphone to focus on the teacher's speech and ignore the other sounds in the room (see Chapter 8). As a result, if this were the only recording you made, you would have a great track of the teacher speaking, but you would lose the other elements in the room, and whether the audience noticed on a conscious level or not, the scene just wouldn't sound (or feel) right. Likewise, if you positioned the microphone to record the ambient sound of the room along with the chalk sound and the teacher's voice all at the same time, the audio quality of all three elements would be reduced to a point that audiences would also notice. The key is to record each element separately, and then to layer the recordings in the Timeline of your sequence.

Layering audio enables you to place each element exactly where you feel it works best. Just as placing the visual elements in a composite sequence

---

SIDEBAR

Layering Audio in *Touch of Evil*

Good soundtracks are layered to create a multifaceted composition, and that includes good soundtracks that use music. Walter Murch layers music with other sounds in a number of films, and he also uses a technique in which he layers different recordings of the same song into a sequence.

In the opening of *Touch of Evil*, the film's star, Charlton Heston, drives through town in a convertible listening to a song on the car's radio. In Murch's version, the song really sounds like it's being played on a car radio. This might seem like a small feat, but if a song were simply added to the soundtrack of a sequence, it wouldn't match the ambient quality of the clips around it. Even if the song had been layered together with ambient sound recorded on location, the song would still sound like the "clean" product of a recoding studio and wouldn't sound like it belonged with the audio recorded on location.

According to a 1998 article in the *SF Weekly* by Michael Sragow, Murch came across instructions from Welles that the music for *Touch of Evil* should be played through the type of speakers his characters used in the film, and then

re-recorded for the film's soundtrack to produce greater realism. The article quotes Welles as writing, "To get the effect we're looking for, it's absolutely vital that this music be played back through a cheap horn in the alley outside the sound building." The technique Welles describes is similar to a technique Murch had already used in a number of films.

"I was dissatisfied with how I heard music that was supposed to be coming from a scene," Murch told the *SF Weekly*. Rather than electronically processing a clean recording to add echoes and other faux imperfections, Murch would actually record the song in its intended environment. "I would just play it in a room, and record it with another tape recorder, so it would have the sound of a radio or a record player in that room." Murch would then add both the clean recording and his re-recorded version to the sequence, and mix them together to form a layered composition. This also allowed him to add other environmental sounds to the audio of a scene to create a deeply textured listening experience.

---

on different layers allows you to position each one independently, placing audio elements on separate tracks affords you complete and much-needed control. Using multiple audio tracks enables you to place each sound exactly where you want it and separately adjust the volume level of each clip. (Positioning audio clips and fine-tuning audio levels are both addressed in detail in Chapter 18.) It may seem both odd and tedious to record all these elements separately, only to combine them at a later stage in postproduction, but it's worth it—especially when you hear the audience ooh and ahh over your work.

SIDEBAR

Sound Perspective, Helicopters, and Wagner

One of the most memorable sequences in *Apocalypse Now* is the Air Cavalry attack led by Robert Duvall. As the helicopters approach the target, Duvall plays a Wagner opera through a set of loudspeakers mounted to the outside of his helicopter.

Not only is the music perfectly suited to the action—a rousingly aggressive opera makes the impending attack that much more powerful and exciting to the audience—but Murch also clearly takes great care to change the sound perspective to match different shots. In film, audio has a perspective just like visuals. When sound is recorded up close, it sounds louder and clearer. When sound is recorded from farther away, it becomes less clear and other audio can be heard along with it. If the action on screen changes from a panoramic wide shot to a tight close-up, the sound perspective should change along with it to maintain realism.

Murch uses a variety of sound perspectives to accompany the action as Coppola cuts from shots inside the helicopters to a series of wide shots showing the helicopters in a fearsome (and somehow beautiful) formation, and then back inside the helicopters again. Although the perspective changes and the sound design incorporates other elements, such as the metallic click of ammunition being loaded into guns and the thundering engines of multiple aircraft flying together, the Wagner score never loses its rhythm; even though Murch cuts between multiple versions of the recording to vary the perspective, he never falls off beat.

The scene builds in visual and musical intensity, until suddenly, the audience finds itself looking at a peaceful and quiet village scene. Everything is quiet and still in the early morning, until we hear the helicopters approaching, and an echoing opera fills the air from far away. Coppola then cuts sharply back to the helicopters flying in over the water, Wagner at full volume and close up.

As the battle builds, the sound design incorporates a climactic section of the opera score, along with explosions, gunfire, radio chatter, and the cries of wounded soldiers. Unlike the opening sequence of the film, which intentionally minimized the connection between the sound design and the action on screen, this sequence uses hyper-realistic audio to bring the audience into the film, making the visuals seem even larger, more intense, and terrifying.

Replacing Missing or Poorly Recorded Audio

The best way to ensure a good sound design for your project is to make sure you record good audio from the start. There are times when you'll need to fabricate audio to match visuals in a project that don't exist in real life, for example, laser gun blasts or spaceship engines. There are also times when you'll need to add sound to strengthen the audio in a sequence, for example, adding footsteps or the wet thud of a punch solidly connecting with its target. This kind of sound addition is easily manageable through the art of Foley sound recreation, explored later in this section. You may also need to record and add voice over narration to a sequence, which is also a reasonably straightforward process covered in this section.

Replacing dialog, however, is a much more complicated business. Creating a single isolated sound and adding it to the appropriate part of a shot, so it looks like an organic part of a film, is fairly simple: look for the first frame in which the action occurs (for example, the previously mentioned fist connecting with an opponent's body part), and then line up the start of your sound effect with the appropriate frame of action in the video track. When the sequence is played back, the sound begins in time with the picture. On-camera dialog is much more difficult to replace, because the sounds have to match the lip movements of the person speaking. A single thud is fairly easy to match with a single punch. Matching full lines of dialog with a person's lip movements is much more complex. (Syncing up separately recorded images and sound is covered in Chapter 8, and placing audio elements in the right spot is covered in Chapter 18.)

Replacing dialog, and why it isn't easy

People don't speak at a constant rate. Even if a person spends considerable time learning exact lines of dialog, he is unlikely to repeat the line at exactly the same speed in a subsequent take. The less experienced an actor is, the less likely he is to match the timing of a line's delivery. Some people are "naturals" at this, and some seasoned actors still have to do many takes to get it right. If you find that the original dialog you've recorded in a shot needs to be redone, the timing of the replacement line becomes especially important, because it needs to fit the lip movements in your shot. If it doesn't match exactly, people notice. And if it's off, it's more noticeable the closer the shot is, and the longer the shot lasts.

Automated dialog replacement

Over time, filmmakers have developed a system for replacing dialog without creating a project that resembles a badly dubbed martial arts film or spaghetti western (production costs are cheaper in Italy, so directors like Sergio Lenone filmed some classic Westerns there using actors who spoke no English and later dubbed in lines of English dialog).

The process of seamlessly replacing dialog is called *automated dialog replacement*, or ADR (also called "looping"), and it's not easy. An actor in a recording studio watches the clip containing the dialog she needs to replace, and re-takes the line while watching herself in the clip. Because actors vary the timing in their delivery of a line, the clip plays in a loop as the actor tries repeated takes to deliver a clean recording that works with the visual footage. A skilled actor can match her own delivery, but it takes great patience and a fair amount of money. ADR facilities are not cheap: they bill by the hour, and as you can imagine, the costs add up quickly. (A day or two of ADR is usually factored into the contract of a Hollywood actor. They are paid extra only if it goes over.)

If you do find yourself recording ADR, save all the takes (you can decide later which to go with) and don't stop recording until you know you've got exactly what you want. If you're going through all the trouble to record replacement dialog, don't compromise by settling for something you're not entirely happy with. Whether you build your own facility or rent time at someone else's, an ADR session is worthwhile only if you don't have to include flawed audio in your finished product. (For suggestions on ways you can record clean audio to begin with, avoiding the need for ADR, see Chapter 8. The chapter also contains a section called "Syncing non-time-code audio" with tips on syncing separately recorded audio and picture.)

> **NOTE**
>
> *Sometimes ADR is simply unavoidable. Much of the dialog of the film* Master and Commander: The Far Side of the World *was recorded using radio microphones, and Director Peter Weir was quoted by* Millimeter *magazine as saying the devices did not fair well in wet conditions (this created particular problems in a film set on an oceangoing ship). "[The microphones] didn't respond that well to humidity or water, so sometimes you only got half of the words, and even then you could only hear one or two over the din."*
>
> *To recreate the lost dialog, Weir and his crew set up large-scale ADR sessions with teams of actors on the set working together. "We had these loop groups of 12 in England, and we'd have them choking each other and fighting to get the mayhem we needed."*

Recording and adding voice over

Adding narration to a project is clearly much easier than adding dialog. Since the sound recording doesn't need to match an exact visual cue on screen, you have much more latitude than when you're attempting to match the lip movements of an actor's dialog. There are a number of different approaches you can employ, and all entail creating a clean voice over recording and blending it with good ambient sound.

> **NOTE**
>
> *If you place the monitor in a room apart from other computer equipment, especially away from noisy hard drives, you'll be able to record replacement dialog without capturing your computer's operating sounds. Some engineers build a soundproof closet to house their computer equipment. I worked with one engineer who simply moved his hard drive to another room and punched a hole in the wall to run the cables through. Whichever route you take, flat screen monitors are very quiet, unlike traditional CRT (cathode ray tube) monitors, which are very noisy, so flat screens make a better choice for ADR use.*

The importance of a clear narration recording

Voice over, or *VO*, is recorded narration that the audience hears without seeing the person who's speaking. It kind of feels like someone is telling you a story. When you record voice over, whether it's narration from the point of view of a filmmaker or one of the characters in a project, you can make the clip much easier to work with by ensuring there's no background sound. If you create a clean voice over recording, you can easily add the narration to the existing sound design of a sequence, and its presence will sound natural to the audience. If instead, your narration recording contains background sound such as a non-matching room tone or a mechanical problem like a 60-cycle hum due to electrical interference or tape hiss that comes from an analog recording device, the additional sound elements can jolt your audience out of the moment, shifting their attention to the change in ambient sound and away from your narration and your story.

One way to conceptualize this idea is to compare narration recorded in a quiet space to a graphic element created on a transparent background, and to compare narration recorded with background noise to a graphic on a random or visually busy background. When you add a graphic on a transparent background, it doesn't distort the material around it, everything else shows through the transparent area undisturbed. Likewise, a clean voice over recording allows the ambient sound of a sequence to play behind it, undisturbed. In contrast, if you were to add narration audio that contained significant ambient sound, it would obscure or color the other audio in the sequence just like adding a layer of busy visuals interferes with the audience's view of everything else.

Effectively blending ambient sound under voice over

Audiences have come to expect great audio along with the compelling images shown to them. If you have background sound throughout your film, but the ambience suddenly drops out to accommodate your narration, the lack of ambience will likely jar your audience out of the story and will probably feel like a mistake.

If you have ambient sound that you recorded along with your images, you can play the ambience in the background to ease the transition into and out of the voice over sequence. For example, if you were creating a documentary about traffic problems in New York City, you could show the audience a shot of stalled traffic and use the accompanying audio you recorded along with the shot. You could then fade the traffic sound down to a lower, but still audible, level, and layer your voice over on top of it. By lowering the ambient sound to a level where it doesn't compete with your voice over, but not eliminating the ambient sound entirely, you ease the transition into and out of the narration. At the end of the voice over, you could then fade the ambience back up to full volume, seamlessly integrating the voice over into your clip.

An introduction to the art of Foley sound creation

Very often, a director will watch an edited sequence and decide that the audio just doesn't carry the weight he is looking for. Material shot in the field might be missing a crucial sound element, or might contain audio that doesn't deliver the effect the director has in mind. At this point, filmmakers turn to *Foley artists*.

A Foley artist creates sound to fit actions in a segment of film or video. Foley creation entails recording audio that sounds like it matches what the audience sees on screen, and then editing that sound into a sequence so it feels organic. A Foley artist works on a stage with a range of floor surfaces and props. A Foley stage also contains a monitor (much like an ADR facility) so the Foley artist can work at creating sounds while watching the footage. Foley artists are commonly employed to create dramatic sounds like broken glass (when glass breaks in real life it often doesn't sound that interesting), gunshots, and other dynamic audio elements. Foley is also used to make us hear sounds that in real life are minor or overlooked, like the rustling of clothing.

Ironically, audiences have gotten so used to Foley sound that non-augmented recordings no longer sound "real" enough to satisfy the discerning ears of the film world. The actual audio "doesn't sound as real as you'd want it to," according to Tony Eckert who mixed Foley sound for *Star Wars Episode I*. As a result, he says on the film's web site, "the artist must find a more suitable object" to create the sound.

Footsteps

One of the most frequent tasks for a Foley recordist is the recreation of footsteps. Using a floor surface chosen to match the ambience of a scene, a Foley artist steps in time with the images on screen and records the sound. A Foley stage commonly contains areas of concrete, wood, tile, gravel, and other surfaces to fit a variety of situations. The Foley artist also varies her footwear to create an appropriate sound (see "Cyborgs, laser blasts, and tall ships" later in this chapter).

To create the sound of footsteps in snow, Foley artists sometimes record themselves squeezing five-pound bags of sugar in time with an actor's footfalls. The crunch of a bag of sugar, especially powdered sugar, sounds remarkably like footsteps in snow—try it out next time you're in the supermarket. (Years ago I was listening to the Steve Allen radio show, and he described working with a new assistant. Allen asked the assistant to get some bags of sugar ready so he could record some footsteps, when he entered the studio, he discovered the assistant had poured the bags of sugar onto the floor so he could step on the contents.)

> **NOTE**
>
> *This is not to say you can never have a section of film without any sound. A Letter To True, a poetic meditation on the fragility of life by filmmaker and fashion photographer Bruce Weber, and* Notre Musique, *the latest film by Jean-Luc Godard, both begin with complete silence and each one literally brought me to the edge of my seat wondering why there was no sound. This is no accident. Weber is an elegant storyteller, and Godard helped create the French New Wave of film when he released* Breathless (À Bout de Souffle) *in 1960. Both directors clearly understand that an unexpected absence of sound can be just as arresting as the most powerful sound effect.*

Cantaloupe and other melons to simulate punches

When it comes time to supplement the sound of a fight scene, Foley artists don't hit actual people, instead they use melons. Somehow hitting thick cuts of meat or dense, heavy fruits such as cantaloupe produces a sound uncomfortably similar to a person getting hit. Sound triggers deeply visceral responses in people, and the audio of a fight scene can be far more disturbing to audiences than the visual depictions of violence on a screen. I have a friend who was okay with watching the onscreen brawls in *Fight Club*, but almost left the theater because she was so disturbed by the realistic impact sounds that went with them.

Breakfast meat as rain

I once worked with an unmotivated sound recordist who wanted to capture the sound of rain, so he stuck his microphone out of a second story window on a rainy day. He was surprised to discover later that he didn't get a good recording. As his technique demonstrates, the sound of rain doesn't come from drops falling through the air. The sound of rain comes from raindrops landing on a surface. I helped my colleague record good rain sound by sitting in my car and recording the sound of rain falling on the car's roof — it produced excellent results.

If they can't get outside during an actual rainstorm, audio engineers sometime simulate the sound of rain by recording the sizzle of bacon. Listened to out of context, the bacon sizzle sounds surprisingly like rain, and vice versa. If you get too many grease-pop sounds it ruins the effect, but otherwise it works nicely.

Cyborgs, laser blasts, and tall ships

As a child, I was quite taken with *Star Wars*, as were all my second grade classmates. I eagerly saw the movie several times, and attentively watched the "making of" special on television, which provided my first introduction to Foley sound. One of the greatest (and most fun) challenges to a sound designer is creating audio elements for objects that don't exist, such as the laser blasts in *Star Wars*. The "making of" special contained a sequence explaining how the sound had been created: effects technicians struck the guy wire of an antenna tower with a hammer. As you can see, well over 25 years later, I'm still impressed.

Foley sound often comes from rather mundane elements put to imaginative uses. To create the purposeful metallic footsteps of the T-1000, Arnold Schwarzenegger's cyborg nemesis in *Terminator 2*, Foley artist Dennie Thorpe wore boots fitted with metallic soles. The same Foley artist worked on *Star Wars Episode I* and, according to the film's web site, realized "those monstrous boots I used in *T2* would work perfectly" to create the sound of *Episode I*'s battle androids. To create the androids' footstep sounds, Thorpe

> **NOTE**
>
> *Various companies offer compilations of recorded effects, such as rain or wind, that you can purchase and use in your work. Generally these recordings won't match the ambient quality of other sound elements in your sequence, but if you layer them together with additional audio elements you've recorded, you can create some very nice ambience.*

wore boots soled with brass, while another Foley artist wore a pair of boots with steel soles.

The sound crew of *Master and Commander* recorded much of their Foley work on the same replica ship where the action was filmed, according to *Millimeter* magazine, even taking it out on days when there were small-craft warnings, in order to get "great thundering crashes and waves smashing against the side of the boat." The crew also rigged a mast in the desert to record sail sounds free of background audio generated by the ocean or the

---

SIDEBAR

The Conversation—Audio That Draws Attention to Itself

In most films, audio is carefully engineered not to shift the audience's attention away from the story. Even in the opening or helicopter attack sequences of *Apocalypse Now,* the audio is not designed as a foreground element for audiences to admire on its own but rather to augment the power of the images on screen. In *The Conversation,* however, audio takes center stage, becoming both a central element of the storyline and of the audience's viewing experience.

The Conversation, Coppola's 1974 film about a professional eavesdropper, follows a highly accomplished audio surveillance expert, played by Gene Hackman, who comes to question his responsibility for what happens to the people he surreptitiously records.

The film centers on Hackman's efforts to record the conversation of a couple walking through San Francisco's Union Square Park. The couple knows they're under surveillance and walks in circles to make it more difficult for any potential eavesdroppers to follow them. Hackman's character, Harry Caul, orchestrates a complex network of undercover audio engineers whom he places in strategic locations around the park (two use giant shotgun mics fitted with telescopic sights). Each recordist captures a different part of the conversation as the couple moves in and out of microphone range, and in a later sequence, Hackman synchronizes and layers the recordings together to form a compete picture.

To dramatize the process of working with audio, sound designer Walter Murch intentionally garbles the sound in a number of sequences so Hackman can refine the audio back into clearly intelligible dialog. In my favorite *Conversation* sequence, Hackman plays the three separate recordings his agents have gathered, each on its own 1/4" reel to reel tape recorder, and manipulates their speeds so they play back in sync. As Hackman increases the speed of one recording and slows down another, the sounds echo

and overlap in response to his adjustments. His actions onscreen immediately impact the complex sound design of the film to the point that it sounds as if Hackman's character is actually creating and manipulating the soundtrack of *The Conversation* while the audience is watching.

In a similarly inventive sequence, Hackman uses a homemade equalizer to filter out sound frequencies and decode a line of dialog central to the film's plot. Before Hackman uses his equalization device, the audience hears the dialog only as unintelligible electronic interference. As he makes careful adjustments on screen, the audience hears the garbled audio evolve into a line of clear dialog that changes the entire meaning of the film. As in the synchronization sequence, the transition from problematic audio to a clear recording highlights just how flawless movie audio generally is, and it also demonstrates how masterfully Murch has intentionally distorted and restored audio to dramatize the professional competence of Hackman's character.

In another sequence, one that I never truly appreciated until I re-watched the film while writing this chapter, Hackman plays the saxophone along with a record spinning on the stereo in his apartment. In true Walter Much form, you can clearly hear two different audio perspectives: the close-up sound of Hackman's on-camera saxophone and the re-recorded phonograph record of his accompaniment (see the sidebar, "Layering Audio in *Touch of Evil,*" earlier in this chapter). The attention to detail in this sequence, and throughout the film, adds depth to Hackman's character and the audience's understanding of the world he lives in. From the sound of an eerie rush of wind in an elevator to the melodious tapping of typewriter keys and the echoing ring of carriage returns (when's the last time you heard that in a film?), or the clicks and whirs of analog recording equipment, this film's sound design doesn't miss a beat.

---

ship itself. This was clearly a group dedicated to recording excellent audio in full detail.

To create battle sounds, supervising sound editor Richard King employed increasingly ambitious techniques. He told *Millimeter* that he "got some crossbow arrows and fired those over mics," then lowered the pitch in post-production to create what he called "this very organic sound." To create "visceral" impact sounds, complete with splintering wood, they spent two or three days firing various pieces of wood with a slingshot—dowels and cut pieces and jagged chunks and other objects—over a microphone.

All this might sound like a lot of work, and it is, but it results in a film that's worth watching. *Hero*, which I mention when discussing color in Chapter 16, makes a similarly ambitious use of audio: there's a scene in which thousands of warriors simultaneously fire arrows, and it sounds like you can hear each one.

Theaters are equipped with increasingly powerful and refined sound systems, and audiences have come to expect detailed, and even thundering, sound design. When your audience hears a well-crafted soundtrack, they'll appreciate it. This doesn't require money, just ingenuity and attention to detail.

I walked out of *Hero* impressed that director Zhang Yimou didn't overlook a single detail, and that impression comes in large part from his use of audio. From the arrow sounds I mention earlier to the perfect use of music in an opening fight scene (not to mention the Itzhak Perlman violin score that holds the entire sound design together), *Hero* is a film that works on all levels. The next chapter explores ways you can use a musical score to augment your sound design without overwhelming the other audio elements in your work. Chapter 18 also addresses the legal issues involved in using recorded music, and introduces software designed to help you create great music for your work without incurring the wrath of someone else's attorney.

Sound Design, from Nuts and Bolts to Fine Tuning

18

The images in the mainTitle sequence you've been working on for the last few chapters have now been carefully structured and refined into a sophisticated, highly polished visual composition. The audio of the sequence, however, is still very basic. There's no sound at all under the opening zoom, or under the title image, and other than a very cute *a cappella* number by Luca, there's a distinct absence of music.

As discussed in detail in Chapter 17, a strong and effective sound design has a tremendous impact on an audience's experience. Augmenting the sound in the mainTitle sequence and molding it into a carefully layered audio composition will result in an infinitely more worthwhile opening for your film.

The following exercises walk you through the finer points of layering sounds, adding music, using audio to bridge a transition from one part of your film to another, and constructing delicately balanced fades to weave individual sound elements into a well-crafted and truly effective sound mix.

Refining the Natural Sound in Your Sequence

As mentioned in Chapter 7, 16 mm and 35 mm film don't record audio. Before the introduction of video cameras capable of recording both picture and sound onto a single tape, images and audio were recorded separately. In some ways, it created an added layer of technical complexity (for instance, an editor had two recordings to work with for each shot instead of just one), but it also freed filmmakers to imaginatively combine the sound from one shot with the visuals from another. One example of how this is done is using natural sound to transition between visual elements.

Natural sound, also referred to as ambient sound, is background audio recorded on location during your shoot (see Chapter 17 for more details). When we recorded an interview of Luca's mom, Deirdre, Luca was singing in the background. Rather than asking him to stop (after all, he's only two years old), we simply recorded him.

In the following exercise, you use the background audio of Luca singing to segue from the main title image to the interview clip in which his mom appears onscreen for the first time. In the opening sequence you've constructed so far, the title image disappears from the screen before we hear any of the audio from the next clip. Sliding the natural background sound of the interview clip to an earlier part of the sequence, so that it begins to play while the title is still onscreen, will create a smoother transition.

> **NOTE**
>
> *To follow along with the exercises in this chapter using Final Cut Pro:*
>
> 1. *Navigate to the Exercises folder on the StartToFinish DVD that shipped with this book.*
>
> 2. *Copy the entire Ch18 folder to your hard drive. The Ch18 folder contains five music files you can use to create a music bed for your opening sequence. The Ch18 folder also contains a reference file (ch18.fcp) that contains a complete version of the opening composite sequence created by following the exercises in this chapter.*
>
> 3. *Open the Final Cut Pro project file you created in Chapter 16 or navigate to the Ch16 folder on the StartToFinish DVD, copy the example file named ch16.fcp to your hard drive, and then open it. The ch16.fcp file contains a complete version of the sequence created by following the examples in Chapter 16.*
>
> *To follow along with the exercises in this chapter using After Effects:*
>
> 1. *Navigate to the Exercises folder on the StartToFinish DVD that shipped with this book.*
>
> 2. *Copy the entire Ch18 folder to your hard drive. The Ch18 folder contains five music files you can use to create a music bed for your opening sequence. The Ch18 folder also contains an example file (ch18.aep) that contains a complete version of the opening composite sequences created by following the exercises in this chapter.*
>
> 3. *Open the After Effects project file you created in Chapter 16 or navigate to the Ch16 folder on the StartToFinish DVD, copy the example file named ch16.aep to your hard drive, and then open it. The ch16.aep file contains a complete version of the sequence created by following the examples in Chapter 16.*

Layering audio and images in Final Cut Pro

This exercise shows you how to move the audio from one clip underneath the visuals of another in the Final Cut Pro Timeline.

1. Click the Linked Selection icon in the top right of the Timeline window. By default, the video and audio of a clip in the Final Cut Pro Timeline are linked, so clicking one automatically selects both.

When you click the icon, it changes from green to black. A green icon means clicking in the Timeline will select related audio and video clips together. A black icon indicates that your selections will no longer be linked, meaning you can now select a clip's audio without selecting the video (and vice versa).

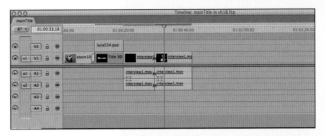

2. Move the Playhead to 01:00:24;12 in the Timeline (you can drag the Playhead or simply enter 01:00:24;12 in the timecode field in the top-left corner of the Timeline). This is the first frame in the clip where an image appears. Although we heard audio in the very first frame of this clip, more than four seconds earlier, the visuals don't start until this point.

3. Click the head of the clip on track V1 to select it. Because the video and audio are no longer linked, Final Cut Pro selects the head of the video without selecting the accompanying audio on tracks A1 and A2.

4. Drag the head of the video clip on track V1 to the Playhead's current location. The video and audio of the clip now begin at different points in the Timeline.

5. Shift-click all the audio clips on tracks A1 and A2, and the second half of the *interview01.mov* clip on track V1. Final Cut Pro selects the clips, enabling you to move them together.

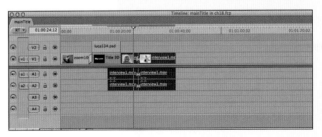

6. Drag the selected clips to the left in the Timeline, until the head of the clip you shortened in step 4 touches the tail of the title image clip on track V1.

The audio of Luca singing (he's improvised his own tune) now starts about half way through the title image clip, and provides an organic transition to the next shot. This type of transition where the video of a clip carries over into the audio of the upcoming clip is called a J-cut, because the shape of the clips in the Timeline resembles a letter J. In cases where the audio of a clip carries over into the video of the upcoming clip, it's called an L-cut because the shape of the clips in the Timeline resembles a letter L (some people use the term L-cut to refer to both L-cuts and J-cuts).

7. Save your work.

NOTE

You may need to render your sequence before you can play it back. If so, follow the steps described in Chapter 11.

Using keyframes in Final Cut Pro to refine your sound mix

The transition you created in the last exercise results in a much smoother sequence. Instead of a hard cut from one shot to the next, the L-cut signals to the audience that a new shot is coming and eases them into the next set of visuals that appears onscreen. You can make the transition even smoother by using keyframes to fade in the audio as well.

To use keyframes to create an audio fade in Final Cut Pro:

1. Click the Toggle Clip Overlays control at the bottom left of the Timeline window. Final Cut displays the keyframe overlays for the video and audio clips in your sequence. The audio keyframe overlay appears as a thin pink line that runs through each of your audio clips.

Just as you used keyframes and the Opacity control to fade a still image of Luca's face onto the screen in Chapter 15, you use keyframes and the audio keyframe overlay in this exercise to fade audio into your sound design.

2. Move the Playhead to 1:00:18;18, which is three seconds after Luca's song first appears in the Timeline.

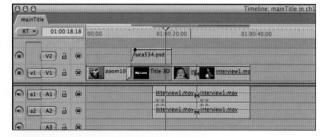

3. Option-click the point where the Playhead intersects the audio keyframe overlay on either track A1 or A2. Final Cut Pro adds an audio keyframe to both audio tracks (because the clip on tracks A1 and A2 contains stereo audio, the changes you make to one track automatically apply to both).

4. Option-click the clip's audio keyframe overlay at a slightly earlier point in the Timeline. Final Cut Pro creates a new keyframe on both tracks.

5. Drag the audio keyframe you added in step four to the bottom of the track, and then drag it to the first frame of the clip. Placing a keyframe at its lowest possible point in the first frame of the clip silences the audio at the head of the clip. Final Cut Pro gently fades the audio up until it reaches its full volume in the second keyframe three seconds later.

6. Play the sequence to preview your completed transition. Luca's singing now fades in smoothly while the title image is still onscreen, and the audio of the *interview01.mov* clip begins playing at full volume before Deirdre starts answering questions in her recorded interview.

The series of small audio adjustments you made in this exercise and the previous exercise result in a much stronger and much more cohesive opening sequence. Nice job.

7. Save your work.

--- NOTE ---

Depending on the specifics of your project, you could use a recording of room tone (which is simply the recorded background sound of a room) to segue into and out of audio recorded in a noisy environment. This could include an office with a loud air conditioner or a restaurant with a large crowd of people talking in the background. Placing a clip of room tone on a separate audio track and then fading it in or out creates a gentle transition between a clip with a noisy background and clips recorded in more quiet environments.

SIDEBAR

Working with Audio Waveforms

As you refine your sound design, you may find it helpful to display audio waveforms in the Timeline. As described in Chapter 8, an audio waveform is a graphical representation of a recorded sound—the peaks in a waveform represent the loudest parts, the flat areas represent the quiet parts. Working with audio waveforms provides you with visual reference points in your sound files, as well as auditory cues.

1. To display audio waveforms in the Final Cut Pro Timeline, press the Command, Option, and W keys at the same time. Final Cut Pro displays the audio waveforms for all the audio tracks in your Timeline.

2. If you have your track height set to the smallest possible size, Final Cut will not display audio waveforms. For more on adjusting track height in your Timeline, see "Inserting additional audio tracks in Final Cut Pro, and layering music elements" later in this chapter.)

To display the audio waveform for a layer in the After Effects Timeline:

1. Click the layer's twirl-down icon in the Timeline. After Effects displays the Audio controls for the selected layer.

2. Click the Audio twirl-down icon. After Effects displays the Waveform controls for that layer.

3. Click the Waveform twirl-down icon. After Effects displays the audio waveforms for the layer you have selected.

Layering audio and images in After Effects

After Effects works differently with clips in the Timeline than Final Cut Pro. After Effects automatically places each clip on a separate layer in the Timeline, as opposed to Final Cut Pro, which places clips on the same track unless an editor specifies otherwise. Because After Effects places each clip on its own layer, sliding one clip under another becomes a more straightforward process.

This exercise shows you how to slide one clip underneath another in After Effects, and how to position it exactly where you want it.

1. Place the Current Time Indicator at the first frame in the *interview01.mov* clip that contains an image of Luca leaning against a red pillow.

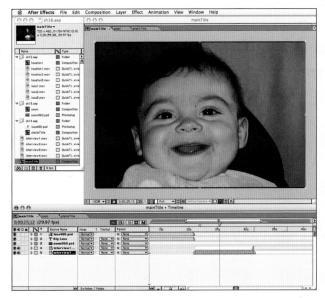

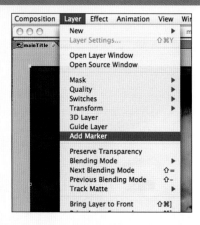

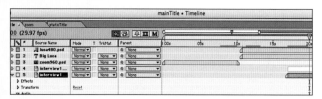

2. Click the *interview01.mov* clip on layer 5 in the Timeline to select it.

3. Select Layer→Add Marker.

After Effects adds a marker to the clip at the Current Time Indicator's present location. In the following steps, you'll reposition the clip to begin playing Luca's audio underneath the image of the main title. Adding a marker to the clip in the frame where Luca's face first appears helps you to keep track of the frame's position in the Timeline as you slide the clip back and forth.

4. Shift-click the second half of the *interview01.mov* clip (on layer 4) so that both are selected. Once both clips are selected, when you move one clip in the Timeline the other clip will move along with it.

5. Drag the clips to the left, so that the marker you added to the clip on layer 5 appears just after the end of the main title image clips on layers 1 and 2.

The audio of Luca's improvised tune now starts about halfway through the title image clip, and provides an organic transition to the next shot.

6. Save your work.

> — N O T E —
>
> *Before you can play your composition to see the changes you've made, you'll need to render your work or preview it using the steps described in Chapter 11.*

Using keyframes in to refine your audio in After Effects

The transition you created in the last exercises results in a much smoother transition from one shot to the next. You can make the cut even smoother by using keyframes to fade Luca's song up to full volume from silence.

To use keyframes to create an audio fade in After Effects:

1. Move the Current Time Indicator about three seconds to the right of the head of the clip on layer 5.

2. Click the twirl-down icon to the left of layer five in the Timeline. After Effects opens the controls for layer 5.

3. Click the Audio twirl-down icon to access the Audio Levels controls.

4. Click the Stopwatch icon for the Audio Levels controls. After Effects adds an audio keyframe to the clip at the Current Time Indicator's location.

5. Move the Current Time Indicator to the head of the clip.

6. Click in the Audio Levels control's numeric field, enter –24, and press return on your keyboard (Mac) or Enter (PC). After Effects lowers the volume level of the clip by 24 decibels (dB), making the head of the clip silent. The volume level then rises evenly until it reaches full volume at the keyframe you added in step four.

You've created a very elegant transition effect in which Luca's singing appears under the main title image and gently fades up to full volume before the visuals from the next shot appear on screen.

7. Save your work.

Adding Music to Your Film

Late one night while watching HBO, I saw a great short film, "Slo-Mo," in which a procrastinating young writer gets stuck in slow motion as the world speeds out of control around him. The film was produced as a graduate thesis project at New York University's Tisch School of the Arts, and uses a number of ingenious devices to create motion effects without a mammoth budget.

To make the protagonist appear visually out of pace with the rest of the world, the filmmakers shot him moving very slowly through crowds of people who were walking at regular speed. When they play the footage back at high speed in various parts of the film, the main character appears to move regularly while people fly past. To complete the effect, when the protagonist was in slow mode, the filmmakers sped up the film's theme song ("Can You Get to That" by Funkadelic) and played the song at normal speed when he was in-step with those around him. Playing the song at a tempo quicker than originally recorded matches the visual effect of people zipping through the frame, and interestingly enough, the song still sounds cool.

The clever use of music in "Slo-Mo" provides a natural way to open the section of this book that explores adding music to your film—the song complements the story and the action onscreen, resulting in a truly memorable experience for the audience. (As you can see, the filmmaker's use of the song was memorable enough for me that I'm writing about it several months later. Coincidentally, as I started the first draft of this chapter, I stopped into

Creating Your Own Music

One way to avoid problems with music copyrights is to create your own. Apple, Adobe, Sony, and other companies make software you can use to assemble prerecorded music clips, called samples or loops, into custom-designed audio compositions. Each application comes with its own collection of samples, and additional samples are available for purchase.

- Apple manufactures two Mac-based products; Soundtrack Pro (*http://www.apple.com/finalcutstudio/soundtrackpro*) and Garage Band (which ships as part of the iLife suite of applications, *http://www.apple.com/ilife/garageband/*). I used Soundtrack to create the music clips supplied with this book.

- Adobe sells Audition (*http://www.adobe.com/products/audition/overview.html*), a PC-based music-creation application designed to work with After Effects, Premiere Pro, and Adobe Encore which is a DVD authoring program.

- Sony Media Software sells Acid Music Studio (*http://mediasoftware.sonypictures.com/products/acidfamily.asp*), which it bills as the "perfect tool for original song creation, and Acid Studio Pro, which offers a greater range of effects. Both products run only on the PC platform.

If you're a musician, you may decide the best way to create music for your film is to compose and record it yourself. If you have friends who are in a band, you might decide to ask them for help with your musical score—just be sure they're not using anyone else's music, and get their permission in writing.

my favorite local café to get an iced coffee, and the guys behind the counter were playing "Can You Get to That." If you can believe it, when I went back to the same café last night after revising a later version of this chapter, they were playing the song again. Apparently, "Can You Get to That" is destined to be a part of my life, or, at the very least, a highlight of this chapter.)

SIDEBAR

Why You Can't Use a Song Without Permission

Some filmmakers try to cut corners by simply copying one of their favorite songs from a CD and pasting it into a sequence, which is problematic for both aesthetic and legal reasons. Even if the audience likes the song as much as the filmmaker does, which is not always the case, when a song is the only element in a sound design it's unlikely to provide a satisfying experience for the viewer. (As discussed in Chapter 17, audiences may find a sequence incomplete if it lacks ambient sound, and music from a CD will rarely match the audio perspective of the visuals onscreen.)

From a legal standpoint, using recorded music without first securing the rights is an especially bad idea. Music is a form of intellectual property—someone created it, and someone owns it. As a result, using music without obtaining permission from the person or company that owns the song may be viewed as a form of theft by a court of law. Owning a CD isn't the same thing as owning a song, music publishers make a distinction between playing a song for your enjoyment at home or in your car, and using it in a movie. (Music publishers also employ large legal staffs to protect their ownership rights, so this is an important distinction.)

Knowing this, inexperienced directors and student filmmakers often decide to use unlicensed music anyway, assuming that few people will ever see their work so intellectual property rights won't be an issue. This creates a problem in itself. When a filmmaker includes material she doesn't have the rights to, exhibition venues may not show the film. Film festivals won't knowingly accept work that contains unauthorized material. If a festival publicly screened the work to a paying audience, the festival organizers could expose themselves to legal problems. As a result, many festival applications require directors to sign a form stating they have the full legal right to use all the material in their films, including music. Broadcast and cable outlets, such as PBS and HBO, will similarly not accept work from a director who can't provide written documentation of permission to use the music in his film.

As you know from your own efforts so far, creating a good film is no easy task—it takes a tremendous investment of time, energy, and more often than not, money. So why create something no one will ever see?

This chapter contains two sidebars that discuss alternatives to illegally using other people's music in your film. "Creating Your Own Music" describes audio production software that enables a filmmaker to create her own musical accompaniment, even if she isn't a musician. "Ownership and fair use" introduces the fair use doctrine, which enables people to use copyrighted material, in very specific circumstances, without obtaining permission.

Using music to enhance, and not overwhelm, your sound design

A successful film score is closely tied to the tone of the project—not just in terms of visual or emotional content, but to the rhythm and pacing of the edits and the transitions from one shot to another. Good music matches, if not furthers, the emotional impact of a film. Speeding up a song is fairly over the top, but then again, so is the entire premise of "Slo-Mo"—when's the last time someone you know got stuck in slow motion? By choosing a style of musical accompaniment that was as silly (but still as well executed) as the visual jokes in the film, director John Krokidas added to the power and impact of his work.

Rather than speeding up or slowing down a song, filmmakers often cut the visuals in a sequence to fit cues in a song. In the nightclub seduction scene from the *25th Hour* (mentioned in Chapter 6 to discuss dolly shots), Spike Lee does a brilliant job of using music to build dramatic tension. The intensity of the music played by the club's DJ grows as the sequence builds to an ill-conceived kiss between high school teacher Phillip Seymour Hoffman and his underage student. Editor Barry Alexander Brown cuts shots of the dance floor in time with the cymbal crashes in the music and the rhythmic scratches added by the DJ. By cutting the picture to fit the music, he creates a tight, tension-filled moment.

However, the music doesn't overtake the sequence. There are other sounds from the club mixed into the soundtrack, and the music is carefully controlled to create a realistic sound environment. For example, the bass thumps at a low volume in the background when Hoffman is inside the club's bathroom awkwardly kissing his student, and the full, loud song becomes clearly audible as he opens the door to leave. As the scene climaxes, the music fades away, clearly signifying (along with a visual change in location), that the film has moved on to a new part of the story.

Even in the opening sequence of *Apocalypse Now Redux* (described in detail in Chapter 17) where music is clearly used as a foreground element, it's not the only sound the audience can hear. Coppola and Murch fade the music in and out along with other audio elements, such as close-up and far away helicopter recordings, to create a complete sound design.

Placing music in your Timeline, and making adjustments

This section of the book walks you through the process of adding music to your Final Cut Pro sequence or After Effects composition. As described earlier in this chapter, simply dropping a song or two into the Timeline of your project is clearly not the most effective use of audio. However, adding music with care and attention to detail can result in an extraordinarily powerful sound design.

> **NOTE**
>
> *You can create a similar effect by using keyframes to play a song at a low volume under one shot and then at full volume under another shot. For more detail on keyframing audio levels, see "Transitioning from your opening theme into the main audio of your film, using Final Cut Pro" and "Transitioning from your opening theme into the main audio of your film, using After Effects" Later in this chapter.*

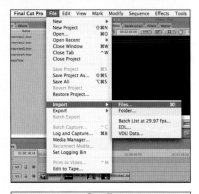

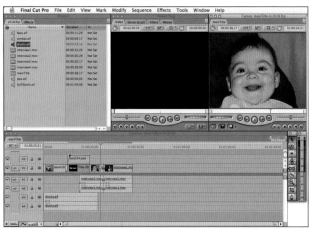

The following exercises show you how to combine and tailor different musical elements to create a polished score for your film.

Adding music to a Final Cut Pro sequence

To add music to a project you created in Final Cut Pro:

1. Select File→Import→Files.... The Choose a File window opens.

2. Navigate to the *Ch18* folder you copied to your hard drive.

3. Shift-click the first file in the window, *bass.aif*, to select it. (.*aif* is a cross-platform audio format, and the five .*aif* files in the *Ch18* folder are different music clips you can add to the project you're creating for this book.)

4. Shift-click the last file in the window, *SciFiSynth.aif*, to select it. Final Cut Pro automatically selects the additional .*aif* files in between.

5. Click Choose. Final Cut Pro imports the file and closes the window.

The imported music files appear in the Browser window.

6. Drag the audio file named *drums.aif* into tracks A3 and A4 of the Timeline, and place it in the first frame of the Timeline window. Because the track contains stereo audio, it takes up two audio tracks.

7. Play your sequence. The drum beat starts immediately in the first frame of the Timeline and continues through the simulated camera movement, the main title graphic, and into the first half of the *interview01.mov* clip where Luca sings and Deirdre begins speaking in voiceover.

By itself, the drum beat is a good start, but like the other elements of your opening sequence, we can refine it to make your film stronger.

Inserting additional audio tracks in Final Cut Pro, and layering music elements

To add additional tracks and additional music:

1. Select Sequence→Insert Tracks.... The Insert Tracks window opens. Adding additional audio tracks to your Sequence enables you to layer additional sound elements over your drum beats.

2. Type 4 in the Audio Tracks field. Creating four new audio tracks will accommodate two additional stereo clips.

3. Leave the After Last Track radio button selected. This will add four new tracks to the Timeline underneath the existing audio in the sequence.

4. Leave 0 in the Video Tracks field. There's no need to add an additional video track to your sequence.

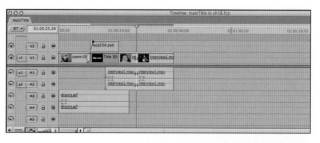

5. Click OK. Final Cut Pro closes the Insert Tracks window and adds four new tracks to the sequence.

The additional tracks may not immediately be visible in the Timeline. To view all eight audio tracks at once, you can adjust the track height.

6. Click the smallest icon in the Toggle Timeline Track Height control at the bottom left of the Timeline window. Final Cut Pro lowers the height of each track so they all fit in the Timeline at once. (Clicking a larger icon in the Toggle Timeline Track Height control would make the height of each track larger.)

7. Position the Playhead at 01:00:05;17 (you can either drag the Playhead or simply type 01:00:05;17 in the Timecode field in the top left of the Timeline window).

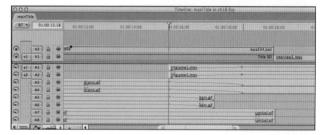

SIDEBAR

Stereo, Mono, and Pan

As discussed earlier in this chapter, a stereo audio file contains two audio clips, one that plays in the left channel and another that plays in the right channel. A mono audio file contains one audio clip, and that clip can be set to play in the left channel, the right channel, or in both channels. When a mono clip plays in both channels, it's called two-channel mono.

You can set the channel that a clip will play in by adjusting its pan control. A clip that's panned to the right or left plays only in that particular channel. A clip with its pan control set to the center plays in both channels.

You can use keyframes to change a clip's pan settings over time. Just as you used keyframes to fade audio clips into and out of the Timeline earlier in this chapter, you can keyframe a clip's pan control to change from one channel to another. You might find this useful if you shoot something that moves from one side of the frame to the other—for example, a person riding a motorcycle—and

you want to adjust your sound to follow along with the movement of the image.

To set the pan control of a clip in After Effects:

1. Click an audio clip in the Timeline to make it the active clip.

2. Select Effect→Audio→Stereo Mixer. The Effect Controls window opens.

3. Enter –100 in the Pan field to pan the clip all the way to the left, +100 to pan the clip all the way to the right, or 0 to pan the clip to the center.

To set the pan control of a clip in Final Cut Pro:

1. Double-click an audio clip in the Timeline. The clip's audio waveform displays in the Viewer window.

2. Enter –1 in the numeric field labeled Pan to pan the clip to the left, +1 to pan the clip to the right, or 0 to pan the clip to the center.

---

8. Drag the audio clip named *bass.aif* from the Browser window into tracks A5 and A6 of the Timeline, and align the head of the clip with the Playhead's current location. Final Cut Pro adds the *bass.aif* clip to the sequence.

The music now builds as the sequence progresses. After a few measures of the drum on its own, a nice bass line begins. In the following steps, you add an additional element of detail by placing a cymbal crash in the frame where Luca's face first appears onscreen.

Adding an audio detail to a Final Cut Pro sequence, and placing it in exactly the right spot

To add, and precisely position, a single audio element to a Final Cut Pro sequence:

1. Position the Playhead at 01:00:10;15. The first frame of the clip containing the image of Luca's face appears at this point in the Timeline on track V2.

2. Drag the clip labeled *cymbal.aif* to tracks A7 and A8, and align the head of the clip with the Playhead's current location.

3. Play the sequence to preview your work. The drum beat starts in the first frame of your opening, and sets a strong energetic tone. The bass line that starts a few seconds later creates a richer, fuller musical score; and the cymbal crash at Luca's first onscreen appearance shows that you don't allow a single opportunity for cinematic innovation to pass you by.

4. Save your work.

Transitioning from your opening theme into the main audio of your film, using Final Cut Pro

The end of the opening theme song that you constructed in the previous exercises now overlaps the beginning of Luca's singing, which you so carefully positioned and faded in earlier in this chapter. The final drum beats overwhelm Luca's voice, drowning out the very nice fade-in you created, and then the drum stops abruptly. In this exercise, you fade out of the drum and bass clips, and then finesse a smooth transition that not only allows the audience to hear Luca start to sing, but unifies all your audio elements to create a gentle and fully nuanced transition.

1. Position the Playhead at 01:00:15;18, which is the first frame of the audio clip on tracks A1 and A2 that contains Luca singing.

2. Option-click the Audio Keyframe Overlay for track A3 at the point where it intersects the Playhead. Final Cut Pro adds audio keyframes to tracks A3 and A4.

3. Option-click the Audio Keyframe Overlay for track A5 at the point where it intersects the Playhead's current location. Final Cut Pro adds additional audio keyframes to tracks A5 and A6.

You've now added keyframes to all four tracks in the music bed at the point in the Timeline where Luca's singing starts to fade in. In the following steps, you add additional keyframes to your music clips and use them to create a crossfade. A crossfade is a type of audio transition involving two clips. As the audio in the first clip fades from full volume to silent, the second clip simultaneously fades from silent to full volume—the volume levels cross during the fade. When properly executed, a crossfade results in a very elegant transition from one sound element to another. In this case, the crossfade you add eases the music out while Luca's singing becomes fully audible—a very professional touch.

1. Option-click near the tail of track A5 to add an additional keyframe after the one you added in step 3. Final Cut Pro adds keyframes to tracks A5 and A6.

2. Drag the keyframe you just placed on track A5 to the bottom-right corner of the track. Final Cut Pro fades out the end of the bass sound.

3. Option click on track A3 directly under the second audio keyframe on tracks A1 and A2. Final Cut Pro places audio keyframes on tracks A3 and A4.

You now have keyframes in your drum track at the same location in the Timeline where Luca's singing reaches its full volume.

4. Drag the keyframes you just added to tracks A3 and A4 to the bottom of the track. This creates a nice, gentle fade that lowers the audio level of the drum clip from full volume to silent at the exact same time Luca's singing fades from completely silent to full volume. The result is a seamless transition from one audio element to another. Your opening title is now fully integrated into the main body of your film, both in terms of picture and audio.

In short, you've created a masterful audio transition. Very nice work.

5. Preview your newly refined sound design, and save your work.

SIDEBAR

Ownership and Fair Use

Copyright law gives people ownership rights to artistic material they produce or purchase. As an artist, you may find that ownership rights help you safeguard your livelihood—people can't use your work without paying for it unless you specifically give them permission. You may also find that copyright restricts your ability to use work created by other people—you may have a great idea that depends on using a song or an image created by another artist that you simply can't afford to pay for.

There's an exception to copyright law known as fair use, which gives artists and academics the right to use copyrighted material without permission, but only under very specific conditions, which are listed below. Courts in the United States have historically recognized the importance of scholarly criticism and the need to use copyrighted material in scholarly work, even if a copyright owner will not grant permission. Filmmakers often claim fair use, thinking it will automatically enable them to use anything they want without paying for it, but in reality, only a limited number of projects will actually fall under fair use protection. What genuinely does and doesn't fall within the boundaries of fair use is subject to considerable interpretation by attorneys and courts of law, so if you plan to make a claim of fair use, be prepared.

There are four considerations that determine fair use, and all four must be met.

- The purpose and character of the use. This asks in what type of project the copyrighted material is being used. For example, if I create a work of scholarly criticism, I can make a stronger argument for fair use than if I create something for commercial entertainment purposes, such as a music video.

- The nature of the copyrighted work. Courts view the use of factual material, such as the information in a newspaper article, differently than the use of an expressive or creative work, such as a song or a painting. If I create a work of scholarly criticism and quote newspaper articles, I probably have a more substantive claim of fair use than if I were using someone else's song in a music video.

- The amount and substantiality of the portion taken, in relation to the copyrighted work as a whole. Using a small section of a copyrighted work, such as a few paragraphs of text out of a 500-page book, may be easier to justify than using an entire song. However, there's no fixed ratio or formula; I've heard people say if you use less than 50 percent of an image or less than two-and-a-half minutes of a song, it's okay, but fair use guidelines don't specify any specific amounts or percentages.

- The effect of the use upon the potential market for, and value of, the work being used. Why buy the cow if you can get the milk for free? If I claim fair use and make someone else's copyrighted work available for free, people may be less likely to purchase the person's original work. As a result, I would have decreased the value of that person's work, which would weaken my claim of fair use.

A good claim of fair use will often hold up in a court of law, even if a copyright owner takes legal action. However, to defend himself against a lawsuit by a copyright holder, a filmmaker must have a solid case. When examining a fair use dispute, a court considers each of the four conditions above. If the court determines a project doesn't meet all four conditions, the filmmaker would not be able to claim fair use.

For more information on fair use and copyright law, take a look at the following:

- *http://fairuse.stanford.edu*: a very thorough examination of copyright law, fair use, and other related issues from the Stanford University Copyright and Fair Use Center.

- *http://www.itvs.org/digitalfutures*: a free guide for media producers, created by the Independent Television Service and the Center for Social Media at American University. This guide, available as a PDF download, provides an "analysis of today's legal, distribution and funding landscape."

Adding music to an After Effects composition

To import music files into After Effects and add them to your composition:

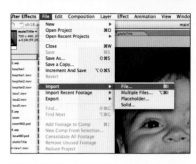

1. Select File→Import→File.... The Import File dialog box opens.

2. Click the first clip in the window, *bass.aif*, to select it.

3. Command-click (Mac) the remaining *.aif* files or Ctrl-click (PC) to select them: *cymbal.aif*, *drums.aif*, *pop.aif*, and *SciFiSynth.aif*.

4. Click Open. After Effects closes the window and imports the files. All five of the *.aif* files now appear in the Project window.

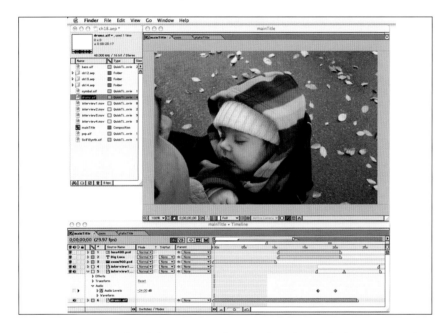

5. Click *drums.aif* and drag it into the Timeline beneath the existing clips at the beginning of the composition. After Effects creates a new layer in the Timeline and places the *drums.aif* clip inside.

NOTE

You can easily preview the audio changes you've made to an After Effects composition by dragging the Current Time Indicator through the Timeline while holding down the Command key (Mac) or the Ctrl key (PC). When you do this, After Effects plays the audio at the Current Time Indicator's location as you move through the sequence.

This technique is called scrubbing, because the action of dragging the Current Time Indicator resembles a scrubbing motion. You can scrub to the right and to the left in the Timeline. When you scrub to the left, After Effects plays your audio in reverse.

Scrubbing enables you to pinpoint a specific audio event, such as a clap slate (described in Chapter 8) or a drum beat. Once you find what you're looking for, you can add a marker to the clip (see "Layering audio and images in After Effects" earlier in this chapter). Adding a marker enables you to easily find the audio event again later. You can also use a marker to visually align a sound element with other elements in your Timeline.

6. Preview your work. The drum beat starts immediately in the first frame of the Timeline and continues through the simulated camera movement, the main title graphic, and into the first half of the *interview01. mov* clip where Luca sings and Deirdre begins speaking in voiceover.

By itself, the drum beat is a good start, but like the other elements of your opening sequence, we can refine it to make your film stronger.

Layering music elements in After Effects

To place additional music clips in your After Effects composition:

1. Position the Current Time Indicator at 01:00:05;17 (you can either drag the Current Time Indicator or simply type 01:00:05;17 in the Timecode field in the top left of the Timeline window).

2. Drag the audio clip named *bass.aif* from the Project window to a location underneath the *drums.aif* clip in the Timeline. After Effects creates a new layer in the Timeline to accommodate the *bass.aif* clip.

3. Align the head of the *bass.aif* clip to the Current Time Indicator's present location.

The music now builds as the sequence progresses. After a few measures of the drum on its own, a nice bass line begins. In the following steps, you add an additional element of detail by placing a cymbal crash in the frame where Luca's face first appears onscreen.

Adding an audio detail to an After Effects composition, and placing it in exactly the right spot

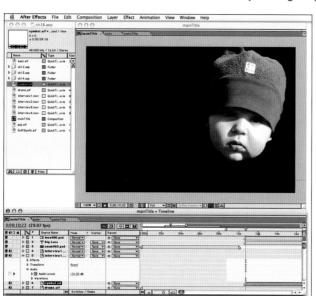

To add, and precisely position, a single audio element in an After Effects composition:

1. Position the Current Time Indicator in the first frame of the clip on layer 1 that contains an edited image of Luca's face. In the following steps, you'll add a cymbal crash to the Timeline so that Luca's first appearance onscreen is accompanied by an audio cue.

2. Drag the *cymbal.aif* clip from the Project window into the Timeline, and position it directly above the *drums.aif* clip. After Effects creates a new layer above the *drums.aif* clip to accommodate the new clip you just added.

3. Align the head of the *cymbal.aif* clip with the Current Time Indicator, and the first frame of the clip on layer 1 that contains an edited still image of Luca.

4. Preview the composition, or render it using the steps described in Chapter 11.

The drum beat starts in the first frame of your opening and sets a strong energetic tone. The bass line that starts a few seconds later creates a richer, fuller musical score. The cymbal crash at Luca's first onscreen appearance shows that you take sound design very seriously and pay careful attention to detail at every layer of your film.

5. Save your work.

Transitioning from your opening theme into the main audio of your film, using After Effects

The end of the opening theme song that you constructed in the previous exercises overlaps the beginning of Luca's singing, which you carefully positioned and faded in earlier in this chapter. The final drum beats overwhelm Luca's voice, drowning out the very nice fade-in you created, and then the drum stops abruptly. In this exercise, you fade out of the drum and bass clips, and then finesse a smooth transition that not only allows the audience to hear Luca start to sing, but brings all your audio elements together to create a gentle and fully nuanced transition.

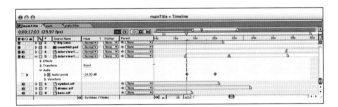

1. Use the vertical scroll bar in the Timeline window to scroll down so you can see all three of the music clips you added in the previous exercises, as well as the first half of the *interview01.mov* clip on layer 5.

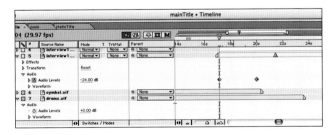

2. Position the Current Time Indicator in the frame where the first audio keyframe appears on layer 5.

3. Click the Layer 7 twirl-down icon to expose the layer's controls.

---- N O T E ----

You can easily preview the audio in a sequence by selecting Composition→Preview→Audio Preview (Here Forward). When you select this option, After Effects plays the audio in your composition, beginning at the location of your Current Time Indicator. If your Current Time Indicator is positioned at the start of your composition, After Effects previews all the audio in your work. If your Current Time Indicator is positioned in a later frame, After Effects plays the remaining audio in your Timeline.

You can also preview the audio in a specific portion of your Timeline by setting the work area to encompass only the section you'd like to preview, and then selecting Composition→Preview→Audio Preview (Work Area). (For more on selecting the work area in an After Effects composition, see Chapter 11.)

You also have the option of rendering your sequence so you can preview the images and audio together. Selecting an Audio Preview is a faster process, so you won't have to wait as long to preview your work.

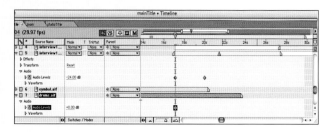

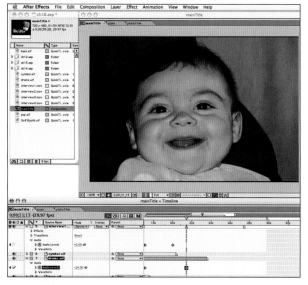

4. Click the Audio twirl-down icon to expose the audio controls for the *drums.aif* clip.

5. Click the Stopwatch icon to the left of Audio Levels. After Effects adds an audio keyframe to the *drums. aif* clip, at the Current Time Indicator's present location.

You now have an audio keyframe in the *drums.aif* clip at the same point in the Timeline where Luca's singing starts to fade in. In the following steps, you'll add an additional keyframe to the *drums.aif* clip and use it to create a crossfade that eases the drum beat out while Luca's singing becomes fully audible—a very professional touch.

6. Move the Current Time Indicator to the first frame in which Luca appears in the *interview01.mov* clip on layer 5.

7. Click in the Audio Levels numeric field on layer 7, enter –24, and press return (Mac) or Enter (PC).

This creates a nice, gentle fade that lowers the audio level of the drum clip from full volume to silent as Luca's singing fades from completely silent to full volume. The *bass.aif* clip ends on its own, leaving the drum beat to fade out evenly as Luca begins singing. The result is a seamless transition from one part of your project to another. Your opening title is now fully integrated into the main body of your film, both in terms of picture and audio.

In short, you've created a very elegant transition by reworking your audio. Excellent job.

8. Preview your newly refined sound design, and save your work.

In this chapter, we took what was a very basic and unsophisticated sound design and cultivated it to produce a highly polished and intricately nuanced audio composition. By repositioning audio clips in the Timeline, adding keyframes to create fades and crossfades, and by layering music clips and positioning each one in exactly the right spot, you raised the quality of your opening sequence to a significantly higher level.

The two remaining sections of this book are appendices covering the process of outputting your film and getting it seen by an audience (watching your work in a room full of people who like it is one of the most gratifying experiences any filmmaker can have). Appendix A covers the creation of a broadcast master, a film print, or DVD; it also explores the fundamentals of streaming your work on the Internet. Appendix B discusses ways you can bring your work to the world, and possibly even make back the costs of producing your film.

This marks the end of the last full chapter of this book. Hopefully, it also marks the start of a very satisfying and productive career for you as a filmmaker. At this point, you've explored digital video from the basic concepts of framing, lighting, and focus, all the way up to the process of developing advanced composite effects, color correcting your work, and producing a professional-quality sound design.

You now have a complete repertoire of technical and creative skills that you can use to create any film you want, and I do mean *any*. By using the techniques explained in this book, and combining them with your artistic creativity and passion as a filmmaker, you can realize any idea you can come up with. If you bring the same creativity and persistence to sharing your completed project with an audience, your film can touch people's lives around the world (see Appendix A).

A journey of a thousand miles begins with a single step, and by learning to use the digital tools available to you as an independent filmmaker, you're well on your way.

As they say in France, *et voilà*—and there it is.

NOTE

If you're inspired to practice working with additional music clips, import the pop.aif and SciFiSynth.aif clips in the Ch18 folder. The pop.aif clip is a cute, cartoon-like sound effect you might want to add when Luca's image pops onto the screen. (You can also try placing it somewhere else in your Timeline. Sound design is, after all, a highly creative process.) The SciFiSynth.aif clip is a very electronic sounding piece that reminds me of the music in the independent science fiction films I remember from my teenage years back in the 1980s. Using this music clip instead of the drum and bass clips I used in the exercises would add a very different feel to your opening sequence.

The Release Print

A

A *release print* is the final edited version of a project that filmmakers show to an audience. The term comes from the days before video when each copy of a film took the form of a print created by a lab. Creating, or striking, a 16 or 35 mm release print was the final step in the postproduction process, and came only after a series of earlier prints had been examined, tinkered with, and finally perfected into a form that was ready to share with the public.

I mention the release print here because even if you're working entirely with digital material, you've reached a similar stage with your project. Whether you decide to create an HD or Digital Betacam master to use in a television broadcast, to strike a film print of your work for distribution to theaters, or to master your work onto DVD or VHS for distribution to people in their homes and in educational settings, you're crafting the finished version of your project. In other words, the last part of this book is about the last step in the postproduction process: creating the equivalent of your release print and getting your work to an audience.

Creating a good film is no easy task. Now that you've gone through the trouble of shooting your dream project and massaging the raw material into a seamless composition, it's time to make the most of what you've got. Appendix A outlines the technical process of creating a final version of your film that you can deliver to audiences around the world, and Appendix B explores different ways you can let the world know your film is out there and worth watching. (You don't need to mount a major advertising campaign for your first film, but it can be very helpful to let the world know that your work is available. Each project you make is a step forward in your career, both in terms of what you learn in the course of production and in terms of establishing a track record for yourself as a filmmaker. For more on some very creative ways independents have brought their work to the attention of audiences, see Appendix B.)

In many ways, these final steps in the filmmaking process are the most important and perhaps the most gratifying. You've made something impor-

In this chapter

Creating a Broadcast Master

Striking a Film Printt Master

Outputting Your Audio

Mastering to DVD, and the Benefits of Distributing Your Work in a Digital Format

Streaming Your Work on the Internet

Making Your Voice Heard

tant, and it's time to share it with an audience. No matter how much you loved making a film, showing it to an appreciative audience and listening to their applause is infinitely more satisfying than working on your project in the isolation of an editing suite.

Finding the Right Postproduction Facility

To find the facility that's right for you, ask around and listen to the advice of people you trust. As I wrote in the sidebar "Finding a Rental House in Your Area" in Chapter 6, the people you know are your greatest resource. If a friend had a good experience at a particular facility, going there yourself might be a good idea. Similarly, if you know someone who had a bad experience somewhere, it might be in your interest to stay away.

Once you find a facility you might want to work with, make an appointment to meet with some of the staff. Bring a preview copy of your project and talk about any technical issues that came up during shooting or postproduction. Judge the responses they give you by not only their apparent technical knowledge, but by their willingness to listen to you and work with you as a client. If they seem like they're not interested in your business, or if they don't appear to take your project seriously, find another place to finish your show. (Remember, as the director of a film, you're the captain of the ship.) The people you work with to master the final version of your film will have a tremendous impact on what the audience sees, so do your best to make sure they're the right people for you and for your project.

A number of the filmmaker support organizations mentioned in Chapter 6 can help you find a good postproduction facility, and may even help you get a discount.

Film Arts Foundation in San Francisco, *http://www. filmarts.org*, has arranged for discounted rates at a variety of post facilities. Film Arts also provides some postproduction services at its own facility. The Association of Independent Video and Filmmakers in New York City, *http://www.aivf.org*, has arranged for its members to receive similar discounts. The Bay Area Video Coalition, *http://www.bavc.org*, also located in San Francisco, offers reduced-cost post services to independents at its own facility. BAVC's services include online video editing (see "Outputting an EDL" later in this chapter) and audio mixing, as well as more specialized services such as closed captioning and technical evaluation of a finished project.

All three organizations are open to membership from people around the world, and many of their discounts are good at places far outside San Francisco or New York. In addition, Film Arts and AIVF both send monthly magazines to their members that may be very helpful to you as a filmmaker (AIVF publishes *The Independent*, and Film Arts publishes the appropriately named *Release Print*).

Creating a Broadcast Master

When you create a program for distribution via broadcast or cable television, the outlet you're working with will likely have very clear technical standards for the work you deliver. This includes the video format in which you submit the final version of your project.

Broadcast outlets often prefer *High Definition* (also known as HD) or Digital Betacam master tapes, because the formats use lower compression ratios than other forms of video, such as a mini DV tape or DVD, and as a result, they deliver superior image quality. Some higher-end film festivals accept only HD or Digital Betacam exhibition copies of a project for the same reason—the festival organizers want to screen the highest quality work for their audiences. (You can almost always still submit a preview copy on DVD or VHS, but if accepted to the festival, you'll need to get them a higher-quality exhibition copy.)

HD is not a single format but rather a collection of high-end video formats that use a variety of frames rates and image sizes. (For more on frame rates, see Chapter 2. For more on the difference between High Definition and standard definition video see Chapter 9.) If you're working with a broadcast organization, or if you've been accepted into a big-name film festival, contact the technical staff as early in the mastering process as possible to find out exactly what they want you to deliver. The organization may have very specific technical requirements, and it's always better to find out what they want before you've spent the time and money to output a final version of your project than after.

SIDEBAR

Negotiating the Right Price

As I wrote in a sidebar about negotiating a price for production equipment rentals in Chapter 6, the card rate that a facility quotes potential customers is often just a starting place for negotiation. Post facilities charge by the hour, and they often charge very different rates for the same services depending on the type of project and the producer's budget.

- Look up the card rate for each facility before you call, so you know their advertised price before you talk to anyone on their staff.

- Always ask for a lower price. A facility only makes a profit if it books customers. Even if you pay less than the advertised rate, the facility will still make more money from your project than if an editing suite sits empty.

- Ask if the facility charges a lower rate to work on projects at night. Large facilities often work on projects 24 hours a day. After the big-budget, high-profile clients go home in the evening, many facilities work on smaller projects at a reduced rate. If you're willing to start your workday after 6 p.m., you may be able to save a significant amount of money. You can also offer to book a session as "bumpable" time, which means the facility schedules you in an open block of time, with the understanding that they many bump your

edit to accommodate a project with a larger budget. If you're patient and have a flexible schedule, this may be a good way to save a few bucks.

- If you're a student, make sure you get a student discount on top of any other price reductions you negotiate. Postproduction facilities offer student discounts for two reasons: they want to attract all the students in a particular program to use their facility (which generates a steady stream of revenue each semester), and they want students to continue to use their facility after graduation (a student who becomes a professional filmmaker may bring all her projects back to the same facility for years to come). If a facility gives you a hard time when you ask for a student discount, it may be a sign that you should take your business somewhere else.

- Ask if the facility will lower your rate in return for a mention in your credits. If your film becomes a success, this could be really good advertising for the facility, and it could save you approximately 10 percent off your bill.

- Ask for an itemized bill, and read it over carefully before you pay. Make sure the bill reflects all the discounts you negotiated, and doesn't include any additional charges.

There are multiple routes you can take to create a professional-quality master tape from your desktop editing system, depending on your source material and your budget. The following pages contain an overview of a few different options. Depending on the route you take, creating a broadcast master can become quite costly. At the same time, since the final broadcast master of your project will be the version of your film that people actually see, this stage may not be the best time to cut corners. Whatever you decide, once your film is finished, you'll have it for the rest of your life. Make this project, and each one you work on, something to look back on proudly.

Outputting an EDL

An EDL is an *edit decision list*, which contains the timecode information for every edit in your project. The EDL is a text document containing:

- The exact source timecode of each audio and video clip in your show

- The name of the source tape where the clip can be found in your original, unedited footage

- The location of each clip in your project's timeline, including the track on which it appears

When you bring an EDL to a postproduction facility, an editor will use the information in your EDL to reassemble a high-quality version of your project directly from your source material. (For more information on timecode, see Chapter 9. For information on organizing your source footage, see Chapter 10.)

For years, creating an EDL and bringing it to a high-end post facility was the only way get a professional-quality master tape from a desktop editing system. Before the introduction of affordable, large-capacity hard drives in the early part of this decade, independent producers routinely worked *offline* to create highly compressed low-resolution versions of their projects on a desktop editing system, and then brought their EDL to a high-end post house. At the post facility, a highly skilled editor would then use the EDL to create the finished *online* version of the project from the original source tapes. This is also called *onlining* or *online editing* and is sometimes referred to as *the online*.

Bringing an uncompressed file to a post facility

Using the steps described in Chapter 11, you can also output an uncompressed version of your project from Final Cut Pro or After Effects and bring the video file on a hard drive to a postproduction house. The post facility staff can then print the material to tape using their equipment. As part of the process, they can also work with you to color correct your footage and further refine your sound mix.

If you decide to bring an uncompressed video file to a post house instead of bringing your source tapes and an EDL, be prepared—uncompressed standard definition video requires approximately 1.4 GB of disk space for one minute of footage. This means that a 30 minute project would require approximately 43.2 GB, and a one hour project would require approximately 86.4 GB.

> **NOTE**
>
> *If you've color corrected your video, added complicated transitions, or created intricate composite effects, the results may not show up in an EDL. If you decide you'd like to finish your film at a high-end postproduction facility, talk to the staff in advance and discuss what parts of the process you should do on your own editing system, and what you should wait to do at their facility.*

SIDEBAR

Outputting an EDL from Final Cut Pro

To output an EDL from Final Cut Pro, select File→Export→ EDL.... The EDL Export Options window opens.

them this, but some very high-tech systems really do read EDLs from a double-density floppy).

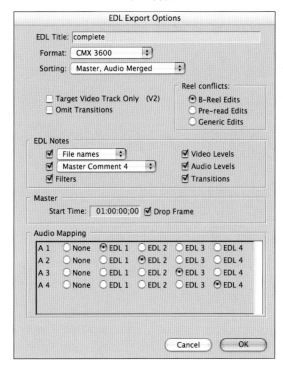

The EDL Export Options window contains an extensive array of settings, almost all of which will vary based on the specific equipment your online facility has available. Before outputting an EDL, contact your post facility and ask them what type of format to use and how you should deliver the list itself. Depending on their equipment, some post facilities may ask you to deliver an EDL on a double-density floppy disk (no one ever believes me when I tell

Because the specifics of exporting an EDL vary to such a tremendous extent, you might also consider outputting a test EDL a few days in advance of your online edit. You can then bring it to your post house along with some source tapes and ask the staff to verify the test EDL using your tape to ensure that everything is okay. If there are problems, the staff can suggest ways to fix them before the online edit.

NOTE

The documentation for Final Cut Pro provides a substantial examination of EDL creation; see "Exporting Edit Decision Lists (EDLs) and Batch Lists" in the Final Cut Pro Help files.

If you're looking for additional information, a book called Video Editing and Post-Production: A Professional Guide *contains a very thorough chapter about on-line editing (http://www.elsevier.com/wps/find/bookdescription.cws_home/675972/description).*

Uncompressed 8-bit HD video requires approximately 7.3 GB for one minute of footage, or 434.4 GB for a one-hour piece. (Depending on the specific HD format you're working with, uncompressed media may require more or less storage space, but HD files by nature require significant amounts of memory. If you have a project that becomes too large to fit on a single external hard drive when you export it as uncompressed media, contact your post facility. The staff may recommend exporting the file in smaller sections that are easier to manage, or they may have other advice depending on the specifics of your project.)

Outputting to tape directly from your system

Depending on the equipment you have access to, and your comfort level with advanced technical operations, you may decide to output a broadcast master yourself, directly from your own system.

Using a digital editing system with a FireWire connection enables you to output media to a compatible deck without a loss in quality. If you invest in additional hardware, such as a high-end bi-directional media converter or video card, you can connect an HD or Digital Betacam deck to your system and output uncompressed master tapes in your own studio. (For more on connecting a deck to your editing system or for details on bi-directional media converters, see Chapter 10.)

As I mentioned in the sidebar "Why You Might Want to Hire a Professional Colorist" in Chapter 16, bringing your project to a professional post facility can provide you with access to specialized equipment outside the prosumer price range. (A high-end HD or Digital Betacam deck can easily cost between 20 and 40 thousand dollars.) More important, going to a good facility provides you with the benefit of the staff's experience and technical knowledge. Even if you have the most highly developed technical chops of anyone you know, it can still help your film to work with a skilled professional specialist—he may see things you don't, or may have a suggestion for improving your work. This is especially true if you consult someone who spends years working 60 hours a week on one specific facet of film production. A good director realizes she doesn't know everything, and surrounds herself with experts in each sub-discipline she needs to work in. This is often referred to as looking at your work "with a second set of eyes." Of course, as the director of the film you don't have to listen to anything that anybody tells you, but it might not hurt to hear what someone else has to say.

Striking a Film Print

Even though an increasing number of exhibition venues will screen video, there are still situations that call for film. Many theaters don't have video projection systems, so depending on where you're screening, you may need to strike a film print to show your work to a theatrical audience. Once you cut together the final edited version of your project, how you get the material from your computer onto a reel of film depends in part on whether you originally shot film or video.

When your source material is video

If your original footage was shot on video, regardless of the video format, you can bring the final product to a post facility for a tape-to-film transfer. During the tape-to-film transfer, the staff of the post house *conforms* the frame rate and aspect ratio of the video master to match the frame rate and aspect ratio of a film print (for more information on frame rate and aspect ratio, see "Choosing a shooting format" in Chapter 2). Depending on the type of video in the source footage, this process can range from straightforward to very complex.

Various forms of HD video use the same 24 frames per second speed of a film print, and a 16×9 aspect ratio that fits nicely in a frame of 35 mm film. In these cases, striking the print is relatively easy, because the frame rate and aspect ratio are compatible from the start. If the frame rate and aspect ratio of the video source don't match those of a film print, the post staff needs to conform the material to fit.

Standard definition NTSC video plays at 29.97 frames per second, as opposed to 24 fps for film. Additionally, NTSC video uses a 4×3 aspect ratio, which is much squarer than the aspect ratio in a frame of film. Higher-end prosumer cameras generally allow a filmmaker to shoot 16×9 video, but only a few models record video at 24 fps, so if you're working with a prosumer camera, you may have to do some conforming to transfer your work to film.

To conform material shot at 29.97 fps to a film-compatible frame rate of 24 fps, technicians put the material through a computer-aided *pulldown* process where the footage is adjusted to play back with fewer frames per second, while not appearing to lose information. To conform squarer shaped 4×3 material to the wider aspect ratio of a 35 mm film frame, technicians may use a *pan & scan* technique to fit a selected area inside the frame. (Pulldowns and pan & scan are both covered in detail in "Choosing a shooting format" in Chapter 2).

As you can see, the process of creating a film print from video footage can be quite involved. Again, this is the last step in your production process, so it's important to ensure the quality of the transfer matches the quality and care you've put into everything else in the project so far. A film print is a big investment, and it's worth it only if you're happy with the results.

When your source material is film

If you shoot 16 or 35 mm film and edit digitally, there are two basic routes you can take to strike a film print:

- The first entails matching the timecode in your edit decision list to the keycode on your negative. Keycode is a series of numbers stamped on the edge of a negative that functions very similar to the way timecode functions in video: using specific keycode numbers, a technician can identify a specific frame of negative, just as he would use a timecode number to identify a frame of video. A professional negative cutter can then cut the original negative to match your final edit. A film lab's staff can then a strike a print from the edited negative.

- The second method entails outputting an HD master of your edited project from the computer you used to edit your footage (see "Creating a broadcast master" earlier in this appendix). You can then color correct the HD master, and with help from the staff of a postproduction facility, you can transfer the footage from video back to film.

Deciding which method to employ depends largely on whether you prefer to color correct in video or in film. Many filmmakers find that digital video, and in particular HD, offers more control over color correction than working in film. Alternatively, some people like to color correct during the process of striking the film print itself.

Creating an HD master for Romantico

Filmmaker Mark Becker shot *Romantico* in 16 mm film and edited his work using Final Cut Pro. *Romantico*, which *filmthreat.com* called "exquisite," follows two undocumented Mexican immigrants who are working in California as mariachi musicians to support their families back home. Becker, who directed the highly successful *Jules at Eight* about an eight-year-old jazz pianist, and also co-edited *The Lost Boys of Sudan*, which follows a group of young Sudanese refuges arriving in the U.S., conceived of *Romantico* as a project that would play in theaters, so to him, shooting film was a natural choice. "If you're thinking about finishing in a theater," he said, "it makes sense to start with as high a resolution as possible," and to Becker, that meant shooting 16 mm film.

Becker had his negatives transferred to uncompressed HD video master tapes as soon as the footage was developed at the lab, and had mini DV sub-masters made with the same timecode. He then archived the HD masters for safe keeping, and imported the mini DV footage into Final Cut Pro to edit.

When he had finished editing his footage, after a total of four years working on the project, he output a DVCAM reference tape from his editing system, then brought it to a post house along with an EDL and his HD source tapes. With the help of the post house staff, he used the EDL to create an HD master from his original film to video transfers. (Bringing a reference tape to an online edit is always a good idea. If there are questions about what a scene should look like, you can show the reference tape to the online editor.)

He then screened the completed HD master at the 2005 Sundance film festival, where *Romantico* made its premiere.

SIDEBAR

Shooting Film and Editing Digitally

Mark Becker started shooting *Romantico* at a time when many independent filmmakers had completely abandoned film and were working exclusively in digital video. Becker planned on editing in Final Cut Pro and had access to a good quality DV camera when he started shooting, but decided from the beginning that he would work in film. In an interview for this book, he told me he made his choice partly for aesthetic reasons and added that "part of what inspires me about shooting film is the way film influences the process." Because film is expensive to shoot and develop, Becker says working in 16 mm leads him to shoot conservatively, and to envision the structure and storyline of a film as he's shooting, instead of trying to capture every single thing that happens. The cost and discipline of working in film, Becker says, lead to a more thoughtful style of production that "helps me take the film to a higher level of what I want to achieve."

Jan Krawitz, who is a highly accomplished filmmaker and a professor in the graduate Documentary Film and Video Program at Stanford University, speaks similarly about the use of film in her own work. The nature of film, Krawitz says, also leads her to shoot conservatively, which "forces me to have a very clear intention for every scene I shoot. The directing occurs in both the shooting and editing stage of the project. People working in DV will sometimes use the camera like a fire hose and only figure out what the film is about once they have to sift through a morass of material in the editing stage."

In 2004, Krawitz completed *Big Enough,* a documentary that revisited five dwarfs who appeared in her 1982 film, *Little People.* Both films were shot in 16 mm and edited on film using a flatbed editing system (a mechanical device the size of a kitchen table, that enables filmmakers to splice shots together from several reels of film and then play the results).

During the shooting and editing, Krawitz said, working in film as opposed to digital video was not a problem. As she neared the end of postproduction, however, she encountered difficulties working with a composer and a title designer who were working with digital equipment, while she was working strictly in film. She also found it difficult and costly to create work-in-progress preview tapes to share with potential funders (a work-in-progress tape is now a standard part of many grant applications, while in the years before digital editing became the standard, people rarely asked for one). Looking back, Krawitz says she would probably edit in Final Cut Pro but still shoot on film. She also sees a definite place for film in the classroom.

"I still do have a bias toward film as a training medium," Krawitz said. (She trained Becker to shoot film at Stanford, and she also trained me in the same program.) "I think if people begin in film, and start to think of an image or a shot as a discrete choice, and don't automatically start in 'sync' sound, which is a given with video, they will possibly think more creatively about the artful potential of the medium."

Becker echoes this sentiment, explaining that his "personal preference" is biased "toward the visual quality of a motion picture film frame," as opposed to a frame of video (a number of other filmmakers, including Daniel Baer, who I quote in Chapter 1 and who shot the footage for *Big Luca,* say very similar things). Like Krawitz, Becker appreciates the flexibility of digital editing, both in terms of the ease with which it enables a filmmaker to experiment with different versions of a sequence, and the ability to return to an earlier version of a project.

Becker, who recently became a father for the first time, also values digital editing because "You can do it in the corner of your living room without bothering your family."

Outputting Your Audio

As you get your project ready to share with an audience, you have a number of audio options at your disposal, ranging from creating a stereo or two-channel mono mix to significantly more elaborate multi-channel mixes. Depending on the equipment you have access to, as well as your comfort and expertise with technical processes, you can output your final audio mix on your own or bring your audio to a postproduction facility.

This section outlines some of the options available to you.

For more information on exporting audio, see "About Multi-Track and OMF Audio Export" in the Final Cut Pro Help files, or "Rendering a Movie" and "Rendering to OMF" in the After Effects Help files

Stereo

As described in Chapter 18, stereo audio uses two channels to produce what's called a *stereo image*: some sounds are sent to the left channel, others to the right channel, and some sounds are sent to both. Regardless of how many tracks you have in your project, you can mix your entire sound design down to two-channel stereo by using the pan controls to assign specific sounds to a particular channel. (For more information on using pan controls, see Chapter 18.)

Filmmakers often create a stereo mix for a program because it works on various types of audio systems. Just about all new televisions play stereo sound, and a stereo mix will also be compatible with newer surround-sound and home theater systems, as well as older, mono television sets.

Creating a stereo mix with the help of a post facility

Many video production facilities have audio engineers on staff, so if you finish your project at a full-service post house, the staff can help you create a stereo mix for your film.

As mentioned earlier in this appendix, an EDL contains information about the video *and* audio clips in your project. When an online editor creates a broadcast master from your original source tapes, that master contains the audio clips you used in the Timeline of your digital editing system. After the broadcast master has been assembled, an audio specialist can then fine-tune the sound clips into a highly refined stereo mix.

An uncompressed QuickTime version of your film can contain stereo or mono audio. In addition, you can output each audio track in your sequence as a separate *.aif* file (for example, if you have eight unmixed tracks in your Timeline, you can export eight separate *.aif* files).If you output multiple *.aif* audio tracks along with your uncompressed video file, an engineer at the postproduction facility can refine them into a carefully crafted stereo mix. (For more details on exporting uncompressed QuickTime files with audio, see Chapter 11.)

When you book time at a post facility, talk to the staff about the technical issues related to your sound design. They may have a particular file type they'd like you to bring to your online session and sound mix, such as an OMF, described in the sidebar "ProTools, OMFs, and your final mix."

Creating a stereo mix yourself

Many video editing and compositing programs, including Final Cut Pro and After Effects, enable filmmakers to create a stereo mix and output their

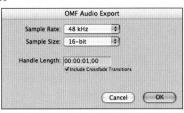

SIDEBAR

ProTools, OMFs, and Your Final Mix

For many years, ProTools has been the industry standard in audio editing, occupying a niche similar to the one Photoshop occupies in the world of still images. Filmmakers very often work with a ProTools engineer to sweeten their audio mix, which means clean up any imperfections and fine-tune all the details in a sound design.

Traditionally, audio sweetening and the final sound mix have been the last steps in the filmmaking process. During an online edit, an editor at the postproduction facility would pull each shot from the original source footage and assemble the video and unmixed audio into an edited master tape. The filmmaker would then bring the edited master tape to an audio engineer who would create the final audio mix, and would also clean up any imperfections in the source audio.

As addressed in Chapter 18, digital video applications such as Final Cut Pro and Premiere Pro now enable an editor to create complex sound mixes as they're putting a sequence together. Even a highly skilled editor can benefit from the help of a good audio engineer, and many bring their work to an audio production facility for that reason. Final Cut Pro allows an editor to export an audio mix as an open media framework (OMF) file, which an audio engineer can then import directly into ProTools. An OMF export outputs a film's sound mix as individual clips, which the engineer can then adjust and rearrange in ProTools as easily you can rearrange clips in the Final Cut Pro Timeline.

To export OMF audio from Final Cut Pro:

1. Select File→Export→Audio to OMF. The Audio OMF Export window opens.

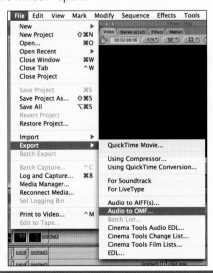

2. Select the sample rate of your project from the Sample Rate pull-down menu.

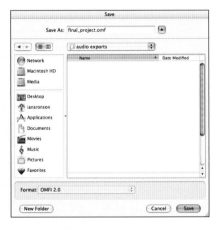

3. Select the sample size, or bit rate, of your audio from the Sample Size pull-down menu.

4. Enter a numeric value in the Handle Length field or you can accept the default length of 1 second. A handle is an "extra" piece of audio that comes before the start of a clip in the Timeline or after the tail of a clip. An audience won't actually hear the handles you include in an OMF export, but they give your editor added flexibility. For example, if an audio editor wants to extend the length of a fade, she'll need extra audio from the handles. Different audio editors will have different handle lengths they'd like you to provide, so contact your editor in advance.

5. Click OK. Final Cut Pro closes the OMF Audio Export window and opens the Save window.

6. Navigate to a location on your hard drive where you'd like to save the OMF. In the example, I've selected a folder named audio exports.

7. Click Save. Final Cut Pro exports an OMF file to the location you've selected. (If you don't have OMF-compatible software installed on your system, such as Pro Tools, you won't be able to open the exported OMF.)

audio and video together, directly from their desktops. For the specifics of setting the pan control of an audio clip, which determines which channel the sound will play on, see the sidebar "Stereo, Mono, and Pan" in Chapter 18. For details on outputting an uncompressed QuickTime version of your film that contains audio, see Chapter 11.

5.1 or 6.1 surround sound

If you plan on exhibiting your work in theaters with high-end sound systems, or distributing your work via DVD to people with sophisticated home theater audio systems, you might want to explore surround sound as an option. While stereo sound uses two channels, surround sound uses many more channels of audio to place sounds in specific locations to the left or right of the listener, as well as behind or directly in front of him.

A *5.1 channel* surround-sound mix uses six channels: left and right as in a stereo mix, plus a left rear channel, a right rear channel, a center channel that faces the viewer, and a sixth channel that plays only low-frequency sounds. Because elements in the mix are directed to different speakers around the room, it sounds like the audio is surrounding the listener— hence, the name surround sound. The mix is referred to as 5.1, even though it contains six channels, because the low-frequency or *subwoofer* channel plays only sounds in the lowest 10 percent of the audible spectrum. Recent developments in audio technology have also produced *6.1* surround sound, which uses seven audio channels.

Creating sound mixes that use more than two channels requires hardware and software beyond what ships with Final Cut Pro or After Effects. Off-the-shelf Macs, and many bi-directional media converters, ship with stereo outputs, as do many new PCs. If you want to work in surround-sound audio you need hardware with at least six outputs, one for each channel, just to monitor the results of your mix. Stereo speakers won't let you hear what a multi-channel mix really sounds like, because stereo equipment simply doesn't output enough channels. A media converter such as Aja's IO ships with eight outputs, and you can send each one to a different speaker. (For more information on media converters, see "Connecting a monitor to your system" in Chapter 10.)

In addition to hardware requirements for listening to your output, working in surround sound requires specialized software that lets you assign different tracks to six or more different channels. Final Cut Pro accommodates only two channels, left and right, but more specific audio production applications such as DigiDesign's ProTools or Apple's Logic Pro are far more multi-channel friendly (the current version of Logic Pro supports surround mixes of up to eight channels).

If you really want to add a surround-sound mix to your project, but don't want to invest in the equipment or take the time to learn all the fine points

> **NOTE**
>
> *If you create a 5.1 channel sound mix and play it back on a stereo system, the system will down-mix the audio to play as two-channel stereo. This means the system will combine all of the channels into just two. If you're working in six-channel audio, it might be a good idea to play your audio on a basic stereo system to hear what a downmixed version might sound like. Audio engineers who mix on fantastically clear, state-of-the-art speakers will often also have a low-end sound system (cheap speakers or even a boom box) in their studio to judge what their work will sound like if someone plays it back on poor-quality equipment.*

of using it, you can bring your work to an audio postproduction facility and have the staff create a multi-channel surround mix for you. As I mentioned in earlier sections of this book, working with an expert may help bring your work to a higher level of quality.

Mastering to DVD, and the Benefits of Distributing Your Work in a Digital Format

Back in the old days (meaning before 2003, when I bought my first DVD burner–equipped editing system), I found it very frustrating to create a work of stunning picture and audio quality using entirely digital tools, and then distribute lower quality copies on VHS tape. Working in an all-digital domain enabled me to edit without a generational loss in quality, but no matter how good the transfer, making VHS copies of a film always degraded my sound and my images. Even after affordable digital cameras and editing systems became widely available in the late 1990s, the most common way to distribute copies of a film remained the unflattering VHS tape.

Fortunately, we now have other options. Computers in the prosumer price range routinely ship with DVD burners that enable an editor to affordably output a finished product in a very high-quality digital format. Newer versions of Final Cut Pro even ship with a program called Compressor, which enables an editor to *compress* a sequence into a DVD-compatible media format directly from the Final Cut Pro Timeline (earlier software applications required outputting a project as a QuickTime file and then compressing the file using a separate application such as Cleaner from Discrete software).

Not only is the quality of a DVD far superior to that of a VHS tape, but DVD production is far more cost effective. Purchased in bulk, recordable DVDs sell at continually lower prices—I've recently seen packages of 100 for $36, and by the time you read this, I imagine prices may be even lower. Many DVDs available at this price are also *inkjet printable*, which means you can design a very professional looking label for the face of the DVD and print it directly onto the disc using a compatible, and often inexpensive, inkjet printer. (Printing directly onto the face of a DVD is much more reliable than using an adhesive label. DVDs spin so quickly during playback that if the label is applied even slightly off center, it can knock the disc off balance, causing problems.) You can then use the same printer to create a professional-looking cover for your DVD case, and when you purchase cases in bulk, you can get them for 30 cents each. When you put all this together, it means you can create a very professional product for less than $1 a copy, including a case and a label.

I'm continually impressed with the quality of DVDs that people make using readily available computer equipment. Work shot with a digital camera and output as a DVD can look surprisingly professional on even a shoestring budget.

Why the VHS Is Still a Viable Option

This section of the book is clearly not a love story between the author and the VHS. However, creating VHS copies of your film can make it much easier to get your work to particular segments of the public. Believe it or not, there are still many people who don't own a DVD player. Rather than exclude a potentially large number of viewers, making a project available on VHS may help you reach more people. Depending on the target audience you're trying to reach—for example, older people who are uncomfortable with new technologies or schools that don't have the funds to update equipment—making your film available in VHS may be a very practical decision.

I've often said that a person's creative ability is more important than the money she spends on equipment. When you make a good project and distribute it to audiences on a well crafted and highly affordable DVD, you can let your creativity shine without going broke.

The DVD production process

There are three stages to making a DVD version of your finished film: compressing your work into a DVD compatible file format, building the DVD architecture, and making copies of the finished DVD. Like outputting a broadcast master and finishing an audio mix, you can complete all three stages of the DVD process in your own studio or you can work with the professional staff of a postproduction facility.

Two popular DVD-authoring applications are Apple's DVD Studio Pro for the Mac (http://www.apple.com/dvdstudiopro/) and Adobe Encore DVD for the PC (http://www.adobe.com/products/encore/). Using one of these programs, you can complete a professional quality DVD from start to finish. For more information on both applications, O'Reilly Media offers two books—Adobe Encore DVD: In the Studio and DVD Studio Pro 3: In the Studio.

This section outlines the process of making a DVD.

Encoding

DVD players read a type of video file called MPEG-2. Once you've finalized all the elements of your film, you can encode the file into an MPEG-2 using a variety of software programs. DVDs also use separate files for video and audio. This enables viewers to choose alternate language tracks and director's and actors' commentary. You can encode video and audio using programs like DVD Studio Pro and Encore DVD. (You can also encode your video and audio using Compressor, which ships with Final Cut Pro, before importing your footage in to DVD Studio Pro.)

Creating a DVD requires filmmakers to compress their material so it fits in the limited storage capacity of a disc. A DVD holds 4.7 GB of information, which amounts to less than 20 minutes of DV-quality video. To fit more information on a DVD, such as a two-hour movie and a collection of special features, video compression programs remove redundancies in a file—information, that in theory, audiences won't notice. However, if video files are over-compressed, they quickly lose their visual quality: blocky compression artifacts can appear in areas of fine detail or at the edges of an image. The trick is to keep enough information so that your film looks and sounds good, but to get rid of enough information so that your project fits on a disc. (For more information on video compression and compression artifacts, see Chapters 2 and 11.)

In addition, not all parts of a film require the same amount of information in order to play back at high quality. Shots with more movement and detail require more information than shots containing less movement or detail.

The amount of information a device is required to read off a disc each second in order to play it back is called the *bit rate*. Current DVD compression programs offer a choice of bit rate settings. The choices vary according to the particular application you're using, but they essentially break down as follows:

One-pass encoding a constant bit rate

The application reads through a project and creates an MPEG-2 version of the video frame by frame, using the same bit rate settings from start to finish. Of the three software-based methods described in this section, this is often the fastest way to create an MPEG-2. (If you decide not to use software-based encoding, you can also use a hardware-based MPEG encoder that converts video files in real time.)

One-pass encoding at a variable bit rate, or one-pass VBR

Unlike constant bit rate encoding, VBR uses a different bit rate depending on the visual complexity of the material. Shots with more motion and areas of finer visual detail are compressed at a higher bit rate than areas with less motion and less detail. This saves space on a DVD by encoding material at lower bit rates whenever possible, and maintains visual quality by encoding material at higher bit rates when needed.

Two-pass encoding at a variable bit rate, or two-pass VBR

Two-pass encoding means the application reads through a project twice. On the first pass, the application analyzes the material to see which frames should be encoded at higher bit rates and which frames can be encoded at lower rates. On the second pass, the application does the actual encoding. Working in two passes enables the software to precisely determine which sections of the video should receive different compression rates, and to adjust the compression applied to each frame accordingly. Two-pass VBR generally produces the best quality MPEG-2 files and makes the most efficient use of disk space. Because it involves two passes through each frame of a project, it's also generally the slowest way to create an MPEG-2.

Building the DVD

Once you've compressed your video, the next step is to build the DVD architecture. This process is called *authoring* the DVD. This entails adding a series of compressed video files to the DVD as *tracks*, which users can navigate to using a series of *menus*. The specifics of how to build a DVD vary considerably depending on the software you're using, but all DVD construction involves creating menus and tracks.

- Menus feature a variety of buttons that you can position and design according to your artistic sensibility and the needs of your project. A viewer can then use these buttons to navigate to different tracks on your disc. When you create a menu you can also add still images, or moving images, that appear in the background, as well as sounds that play while the menu is onscreen.

- Tracks are MPEG-2 files that will play in the order you specify. You can set a track on the DVD to play automatically when a viewer inserts the disc in a DVD player, or you can set the DVD so each track plays only when a viewer selects a specific menu option. You can also add chapter markers that enable a viewer to navigate to different points within a track.

Copying the disc

There are two ways to produce copies of a DVD: *duplication* and *replication*. Although they may sound similar, they're actually very different.

- Duplication simply means making duplicate copies of a DVD, which you can do yourself if you have an editing system equipped with a DVD burner. Apple's DVD Studio Pro and Adobe Encore both enable you to build and burn DVDs from your own computer, as do many other DVD creation programs. Properly made DVDs produced using one of these applications will play back on almost all DVD players, although some older models may have trouble reading a disc burned by a computer instead of a replication facility (replication is described below). Once you've created a master disc, you can easily create copies using a program like Roxio Toast (*http://www.roxio.com/en/products/toast/*). If you have an inkjet printer, you can design custom labels and covers for the discs you burn and quickly create professional quality DVDs to sell. The only catch is, these discs are not only easy for you to copy, anyone with a computer and a DVD burner can make copies as well.

> —— NOTE ——
>
> *Always test each DVD on a standalone DVD player connected to a television (also called a set-top DVD player) to make sure everything works the way you want it to. Occasionally, a DVD you burn yourself will play back perfectly on a computer but won't look right when you play it back on a set-top player (or vice versa). Other times, something might go wrong with the burn, and your DVD will simply fail to play back all together.*
>
> *If you encounter problems, try restarting your computer and burning another copy of the disc. (Turning a computer off, waiting a few seconds, and then turning it on again can solve a surprising number of technical problems.) If you burn another DVD and find you still have trouble playing the DVD back on a set-top box, try re-encoding your material at a lower bit rate. Computers can sometimes play tracks encoded at higher bit rates than a standalone DVD player can handle.*
>
> *Some DVD players that are more than five years old cannot play discs that you burn on a computer, but will play only discs created at a commercial replication facility.*

If you don't want people making their own copies of your work, you need to bring your work to a replication facility to add copy-protection features.

- At a replication facility, technicians create a glass master of your DVD and then use the master to create additional copies. A large facility can easily produce hundreds or even thousands of copies of a DVD at once. Creating a glass master is a fairly expensive process, and becomes cost effective only when you produce more than 1,000 DVDs. (At the same time, if you're using DVD Studio Pro, it's the only way you can add copy-protection features designed to prevent people from making duplicate VHS or unauthorized DVD copies of your work. If you're using another DVD authoring application, the software may enable you to add copy protection, but these features may produce discs that will not play on all systems.)

The specifics of the replication process will vary depending on the facility you choose to work with. Contact the facility in advance to find out about particular technical requirements and how they would like you to deliver your material. Some replication facilities may accept a DVD that you burn from your computer, others may require you to deliver a project on a Digital Linear Tape, or DLT. Not all DVD-authoring applications are capable of outputting to a DLT, so be sure the software you're working with is compatible with the replication facility's delivery requirements. If you need to use a DLT, contact one of the organizations listed in the sidebar "Finding the Right Postproduction Facility" and ask if they can recommend somewhere in your area to rent one. You can also ask the replication facility if they can recommend a rental service near you.

> **NOTE**
>
> *As mentioned earlier in this appendix, avoid using adhesive labels on your DVDs. In addition to causing problems during playback, a paper label can get stuck in a DVD player.*
>
> *On commercially replicated DVDs artwork is silkscreened onto the face of the disc. If you'd like to create customized artwork for DVDs you burn yourself, you can purchase inkjet printable DVDs. These DVDs can be inserted into a compatible inkjet printer, which prints directly onto the face of the disc and can produce professional looking results.*
>
> *The Epson Stylus Photo series of desktop printers (http://www. epson.com/) prints directly onto both DVDs and CDs.*

Streaming Your Work on the Internet

Streaming your work from a web site means your video plays while it's loading to your viewer's computer. Instead of having to wait until the entire file has downloaded, which when you're working with video files can be quite some time, streaming delivers a continuous feed of information that enables your viewer to play your content without excessive waiting time.

Like mastering tracks for a DVD, streaming requires that you encode your work into a format compatible with the system your viewer will be using to play it back. Depending on the technology you use to stream your work (see the section on *codecs* below), the particular encoding software you use and the file types you create will vary. Regardless of the specific technology you're using, however, the basic principles of streaming revolve around *bandwidth* and *processor power*.

- Bandwidth refers to the amount of information a given Internet connection can transmit at any one time. A computer with a high-bandwidth Internet connection (such as a DSL line or cable modem) can receive

— NOTE —

Not all users want to watch video when they come to a web site.

If you're streaming video on the Web, give your users control. Let them click to load, instead of using some type of autoload command in the code of your web page. Also, if it's a really long clip, edit the video into segments and enable the user to stream each part of the clip when he's ready. You can also make your site much more easy to use by clearly labeling the links to each clip and providing a brief description of each clip's contents and its length.

more information than a computer with a low-bandwidth Internet connection (dial-up). As mentioned in the earlier section on encoding DVDs, video files require considerable amounts of information to play back, and if you remove too much information, the quality suffers. Understanding whether your audience will be using high- or low-bandwidth Internet connections is important to you as a producer, because you can then tailor your content accordingly. If you know most of the people watching your work will be connected to the Internet via low-bandwidth modem connections (and many people still do use 56K dial-up modems), you'll need to create files that require less information to play back. One way to accommodate viewers with varying connection speeds is to offer different versions of your work, and allow a viewer to choose. For example, the movie trailers on Apple's web site (*http://www.apple.com/trailers/*) are available in three different sizes: small, medium, and large.

- Processor power refers to how much information a particular computer can process at any one time. Video requires substantial amounts of processor power, and as mentioned in Chapter 10, a slower computer might freeze while playing back a video clip if the computer's processor becomes overwhelmed. When you're distributing work via the Internet, understanding the limitations of processor speed will enable you to make your work accessible to a much wider audience. Even if the people viewing your work have especially fast Internet connections, they won't be able to watch full-screen video on a computer that doesn't have a powerful processor.

Optimizing video for the Web

Optimizing a video file for web streaming means converting it into a form that's particularly well suited for the Internet. The trick to making video files accessible to a large audience of Internet users is to manage both the image size and frame rate.

As you can imagine, a 720×480 video clip requires much more information than the same clip reduced to 360×240. The images would appear noticeably smaller, but they would also require much less bandwidth to stream effectively and much less power to play back. Similarly, a video clip that plays at 30 frames per second will require much more information than the same clip reduced to a frame rate of 15 fps. As noted in Chapter 2, clips played at a slower frame rate may mot look as smooth as clips played at a higher frame rate. The trick is to balance the affordances of the medium with your goals as a filmmaker. If you know you're going to be streaming your work, you can shoot more close-ups (they look good at reduced frame sizes) and shots with fewer pans and zooms (motion within the frame often looks better than camera movements at lower frame rates). If you're editing a trailer for the Web, you might use more close-ups and fewer zooms and pans than you would for a trailer meant for TV or theaters.

Codecs

The term *codec* is short for compression/decompression. When you apply a codec to a video file, you're applying a compression algorithm that makes the file easier to transmit over the Internet. (A compression algorithm is essentially a complex mathematical formula.) When viewers play your file back, they need to use a compatible codec to decompress the file.

The QuickTime player, which is available for both Mac and PC, uses a variety of codecs and comes preinstalled on all new Macs and many new PCs. If you're using QuickTime Pro (which is available at *http://www.apple.com/quicktime/*) to encode media, you can choose the particular codec you'd like to use. When a viewer plays the clip back, the QuickTime player will automatically choose the appropriate codec to decode your work. You can also use proprietary media players to encode your work, such as Real Player (*http://www.real.com/*) or Macromedia's Flash Player (*http://www.macromedia.com/*), which use their own codecs. Real Player and Flash Player are available as free downloads and are both widely used.

A number of software applications encode video into formats you can stream from a web site. Some widely used applications include:

Discreet cleaner (http://www.discreet.com/products/)
> Available for both Mac and PC, cleaner encodes video and audio using a variety of codecs, enabling you to choose the specific settings that work best for your project. cleaner also encodes media for DVDs and for some mobile devices.

Macromedia Flash (http://www.macromedia.com/software/flash/)
> A streaming media authoring tool available for both the Mac and PC platforms, Flash enables you to encode video and create a customized web-based interface.

Apple Compressor (http://www.apple.com/finalcutpro/compressor.html)
> This product ships with Final Cut Pro and is designed to work with both Final Cut and DVD Studio Pro. Compressor is integrated with Final Cut Pro, so you can export audio and video directly form Final Cut using a variety of codecs.

You can also output video files using a variety of codecs directly from Final Cut Pro and After Effects. See "Exporting a QuickTime Movie" in the Final Cut Pro Help files, and "Exporting footage using QuickTime components" in the After Effects Help files.

Regardless of the software you're using, experiment with different codecs, image sizes, and frame rates to find the settings that work best for your particular footage. Select a few different clips, and encode each clip a few times using a few different settings to see how the material will play back. Depending on the contents of the clip, different compression settings may produce very different results. For a discussion of temporal and spatial compression as they relate to the content of a shot, see "Compression algorithms" and "Shots that make compression easier" in Chapter 3.

NOTE

If you're interested in learning more about Flash or Action-Script, which is the coding language Flash developers use to create interactive online environments, O'Reilly offers a number of books including: Programming Flash Communication Server; ActionScript for Flash MX: The Definitive Guide; Flash Hacks: 100 Industrial-Strength Tips and Tools; ActionScript Cookbook: Solutions and Examples for Flash MX Developers; *and* Essential ActionScript 2.0.

Even though they're each manufactured by separate companies, you can use the applications mentioned above together. For example, I used Compressor and Flash in combination to produce my latest online project, *Mi Querida America*, which follows a group of immigrant teens through their first year of high school in New York City. I optimized my video in Compressor, adjusting the frame rate and image size, and then designed an interactive interface using Macromedia Flash. The results are online at *http://www.digitaldocumentary.org/america/*.

Making Your Voice Heard

Whatever technology you decide to employ, they all lead to the same goal: bringing your work to the greatest number of people. I make a living because I know how to use technology, and on a good day, I even think technology can be fun. What I really love, however, is making something really good and using technology to share it with an audience.

The previous 18 chapters were devoted to making the film you always wanted. This appendix explored various electronic media you can use to deliver that film to an audience, and the next and final section of the book explores ways you can develop the greatest audience possible. Appendix B discusses ways you can make sure people see the great film you've worked so hard to make.

For years I dreamed about having my own production studio because it would enable me to make any film I wanted. Now I have a great digital editing system, and I can use it to make the films I want, in exactly the way that I think is best. Technology provides me with control over both creation and distribution.

When I present my work at conferences and film festivals, I often begin by saying that I don't have enough money to buy a television station, but I do have the technical ability to make a DVD or a web site and deliver my work to anyone in the world. Even if I make the best television show ever, I still have to wait for a television station to broadcast it—and if a programming executive doesn't think my work is worthwhile, that show will never see the light of day. Because I control my own means of digital production, I don't have to wait for someone's permission to exhibit something that's important to me.

And neither do you.

The Last Step: Negotiating a Sale (and Why You Should Always Bargain Up)

B

The three greatest pieces of advice I ever received as a filmmaker are as follows:

- If a piece of equipment isn't working, make sure it's plugged in.
- No film is ever finished, you just stop working on it.
- When it comes to showing your film, any festival is better than no festival.

Each of these lines is entirely applicable to this book, and the last is particularly relevant to this appendix—especially when you get to the point where you've completed a film and have to decide what to do with it. (Incidentally, all three came from Academy Award–nominated documentary filmmaker and cinematographer Jon Else who taught part time in my graduate film program.)

Getting Your Film Seen: How Festival Exposure and Press Coverage Can Get You Noticed

For many independent films, festivals present the best way to find an audience. Film festivals specialize in bringing audiences work they otherwise wouldn't see, so if you've got something really good, entering it in the right film festival is an important first step.

In addition to the big name film festivals, such as Cannes and Sundance, there are hundreds of others all around the world, ranging from very large general audience film festivals to events that cater to distinct niche markets. There are Asian film festivals, African American film festivals, Latino film festivals, and Jewish film festivals across the United States. There are also festivals that focus on films about human rights issues, environmental concerns, gender issues, and themes related to sexual orientation (just about every large- to medium-sized city in the U.S. has a gay and lesbian film festival, and so do many smaller ones).

If you're a student, you can enter your film in student categories (which is great because then you don't have to compete with people who have years more experience than you do). There are also some festivals that take only student work, so if you're eligible, these can be particularly good places to enter your film.

There are festivals devoted exclusively to short films, films by women, films about history, films about science, and pretty much any other category or subcategory you can think of.

In short, whatever your film is about, there's a festival out there with your name on it.

If you get into a big festival, there may be programmers from other festivals in the audience who like your work and invite you to exhibit somewhere else in the future. Festival screenings also give you clout as a filmmaker. Anybody with a camcorder can make a film, but if you exhibit your work, it means someone has given your project a seal of approval. Likewise, if someone reviews your film, it helps draw attention to your work and builds your track record as a director or producer.

Festivals are also places where film distributors look for projects they'd like to acquire. Filmmakers Chris Kentis and Laura Lau made *Open Water* (described at length in Chapter 16) for $130,000 and when the film played at Sundance, sold it to Lion's Gate Entertainment for $2.5 million, according to the *New York Times*. Shopping your film around to various festivals probably won't make you a millionaire, but a respectable list of festival screenings and some good print reviews can make your film much more attractive to a distributor and bring you some much-needed clout when it comes time to negotiate a contract.

The downside of sending your work to film festivals is the process opens you up to rejection. Festivals often receive far more submissions than they have the ability or inclination to exhibit. This means that although some people get accepted to just about every film festival they apply to, most don't. Even people with exceptionally successful films often get rejected from a number of festivals before their film catches on and starts getting shown. In fact, filmmakers often get rejected from one festival and then receive an invitation to a more selective film festival soon after. (It's also not unheard of to get rejected from a festival and then invited to exhibit at that same festival after your film has gotten some good publicity.)

Ultimately, you have to believe in the quality of your own work and your own abilities as an artist. If one selection committee turns you down, find a better one somewhere else.

NOTE

There are entire festivals devoted to films that can be used in an educational context. Classroom use is an important and all too often overlooked venue for filmmakers. According to Larry Daressa of the educational distribution company California Newsreel, http://www.newsreel.org, educational video is a $2 billion a year market.

SIDEBAR

The Business Card Pitch

A pitch is a short summary of your film that you give to a potential funder or distributor. In essence, a pitch is a way of saying, "This is why my project is really good, and why you should give me money for it." (In film, as well as other creative professions, the word "pitch" functions as both a noun and a verb. For example, "here's my pitch," or "I'm going to pitch my project at the meeting."

When you show your work at a film festival or attend a conference, you'll meet lots of people and have lots of opportunities to deliver your pitch. Of course, funders and distributors hear tons of pitches all day long, so to making your pitch both short and memorable can help make it much more effective. Two names for a well-rehearsed and succinct pitch are the business card pitch and the elevator pitch. A business card pitch gets its name because it's short enough to fit on the back of a business card if you write it down. The elevator pitch is closely related, and gets its name from the fact that it's short enough to tell someone during a brief elevator ride.

The pitch for my most recent project, *Mi Querida America,* is "a documentary following a group of immigrant teens through their first year of high school in New York City." The pitch is short, easy to understand, and easy to deliver.

I pitched an earlier project, *Tell Them You're Fine,* as "a documentary following three recently diagnosed cancer patients as they come to terms with their illness."

A good pitch can also include details about the film's target audience. Depending on who I was speaking with, I would often add that *Tell Them You're Fine* "is aimed at cancer patients and their families, as well as hospitals and medical schools." (*Tell Them You're Fine* is distributed by Fanlight Productions: *http://www.fanlight.com/.* I read about Fanlight in the *The AIVF Guide to Film and Video Distributors,* mentioned later in this appendix, called them up because I thought they'd be a good fit for my project, and gave them my pitch over the phone. They liked my pitch enough to view a copy of the film, and liked the film enough to offer me a contract.)

Refining your pitch and learning to deliver it with confidence is an essential skill for any filmmaker. As you go out into the world with an idea for a film (or with a project that's in progress or something you've already completed), your pitch is the best way to open doors. Before funders and distributors can pay attention to you, they have to know you exist, and the pitch is your means of introduction. Make each one count.

How to Find Out About Festivals

Film Arts Foundation lists upcoming festivals in its monthly magazine *Release Print,* and also lists calls for entry in the members area on its web site, *http://www.filmarts.org/.*

The Association of Independent Video and Filmmakers lists upcoming festivals in its monthly publication *The Independent.* (If you're serious about trying to get your film out there, it's a good idea to read both *The Independent* and *Release Print,* they often list different opportunities.)

The International Documentary Association lists upcoming festivals in a public area on its web site, *http://www.documentary.org/.*

The Option of Self-Distribution and the Story of Mary Jane's Not a Virgin Anymore

In 1997, a first-time feature director named Sarah Jacobson screened *Mary Jane's Not a Virgin Anymore* at the Sundance Film Festival. The film explored sex from a woman's perspective and had a strong female lead, making it very different from what audiences had seen in mainstream features.

Although *Mary Jane's Not a Virgin Anymore* was well received at both Sundance and South By Southwest, two very influential film festivals, Jacobson wasn't offered a distribution deal for her film.

"These Hollywood people, these investment bankers said girls don't go to movies unless their boyfriends take them," Jacobson told the *San Francisco Chronicle* in 1998. Not only did she strongly disagree, but she decided to prove them wrong.

Rather than waiting for a distributor to bring her work to the public, she set out to do it herself. Working with her mother, Ruth, Jacobson sent her film to a number of high-profile festivals and then literally became her own distributor, booking screenings at theaters across the United States. The *San Francisco Chronicle* described her daily activity as follows:

> She's on the telephone for hours each day, schmoozing small theater owners across the country. She sends them a tape of her film and badgers them for a commitment to screen the movie. That's when she's not sending out hundreds of mailers, contacting the media and stapling neighborhoods with posters.

"You have to be pushy, you have to be assertive, you have to be a pest," Jacobson told the *Chronicle*. She traveled to each screening, and brought a print of the film with her (this was in the pre-digital era when screening in a theater meant delivering an expensive 16 mm film print to each show). Her success as a filmmaker clearly came from her sheer determination: instead of waiting for someone to offer her an opportunity, she created her own. According to an article in the *Village Voice*, Sarah and Ruth Jacobson "scraped together 50 grand" to make the film "in part by sending postcards to complete strangers asking for money."

I decided to include Jacobson's story in this appendix to inspire you as both an artist and an entrepreneur. Jacobson used her artistic passion to share her work with the world at a time when film production and distribution were far more expensive and labor intensive than they are today. She had a strong vision of what she wanted to share with an audience, and she worked very hard to make sure her vision was fully realized.

Sadly, Jacobson died of cancer in early 2004 at the age of 32. A *Village Voice* article printed shortly after her death described her as someone who "packed more into her heartbreakingly brief lifetime than most do in lives three times as long."

There's a really good interview on *filmvault.com* from 1998, in which Jacobson talks with the *Austin Chronicle* about creating films from a female point of view and what led her to make *Mary Jane's Not a Virgin Anymore*.

SIDEBAR

Your Work, Your Audience, and the Q&A

When you show your film in public, you'll often be invited to talk with the audience afterward. These question and answer sessions, or Q&As, are great because they give you a chance to speak directly with people who've seen your work. They also give you a chance to thank people who helped make your film possible (this includes all the people who appear in your film, your crew, and anyone who gave you money—a few well chosen words of public thanks can go a long way).

On a good day, the room will be filled with people who want to tell you how talented you are and how much they love your film. Unfortunately, there may also be some very annoying people in the audience. I was once at a film festival Q&A in Mexico City where an audience member told a director that he thought the subject of his film was excellent, and then told the director that he thought his filmmaking was abominable and had absolutely ruined what otherwise would have been a very interesting story. The director very calmly thanked the audience member for his comment, and then moved on to discuss other aspects of his film with other less-hostile people. I was impressed with the director's response—rather than further engaging a belligerent person and enabling him to deliver additional negative opinions, the director kept his cool and shifted the discussion else-where.

I recently traveled to Berkeley, California, to present *Mi Querida America* at the International Conference on Technology and Society. The morning of my presentation, I confided to my girlfriend, who had traveled to the conference with me, that I was a little nervous that the audience might not like my work. She asked if I thought my project was good, and when I answered yes, she told me that if I was confident in the quality of my work, it didn't mater if someone in the audience didn't like it. (She also said if people didn't like my work they were stupid, which reconfirmed my opinion that I had found myself a very special woman.)

I mention both these stories here because showing your work in public takes a lot of guts. You're showing something that represents your time, your effort, and your artistic judgments—and sharing something that important can sometimes be intimidating. At the same time, as I've men-tioned before in this book, showing your work to a receptive audience is one of the greatest experiences you can have as a filmmaker. If you believe in yourself, and you believe in your work, get out there and show it to people. If you meet people who don't like what you've come up with, you don't have to listen to them.

If you're interested, take a look. Hopefully, it will inspire you to new heights of success as an artist:*http://www.filmvault.com/filmvault/austin/m/maryjanesnotavirg1.html*

D.I.Y. or Die: Taking Your Work on Tour

In 2002, Michael W. Dean completed his first film, *D.I.Y. or Die: How to Survive as an Independent Artist* (*www.DIYorDie.org*). The film features interviews and performances by artists ranging from Lydia Lunch and Fugazi to GWAR and filmmaker Richard Kern. The cost-effective nature of digital production enabled Dean to create exactly the film he wanted to as an independent producer, without having to answer to anyone else.

In an interview for this book, Dean explained that after he finished his film, he then brought the same sense of enthusiasm and self-determination to making sure his work found an audience. "I sort of made the film my *mission* for about two years. I lived, breathed, and slept that film. I jumped out of bed every morning excited to work on it. If something doesn't make me feel like that, I won't be involved in it."

Independents often bring tremendous passion to their work, only to find that traditional means of distribution, such as a public television broadcast or a large theatrical release, may not be available to them. At this point, all too many filmmakers give up in frustration. Rather than waiting idly and allowing his film to sit on a shelf collecting dust, Dean created his own distribution plan. "Even though I got turned down by PBS, and every indie film 'art' distribution company in America, I took the film on a tour myself, showing it in clubs and bars across America. Then I did the same in Europe." As a result of his self-financed tour, Dean secured a distribution deal that got his DVD into every Tower Records in the country, listed on Netflix, and into many video rental stores. The company Dean works with, Music Video Distributors, found Dean because they saw his web site, and "they looked at the web site because someone at MVD heard about the film because I toured with it in their state."

Dean told me that he did *D.I.Y. or Die* on his own, and continues to do things himself, because he hates waiting for permission to make art. Taking control of production, and making what you think is important, is in many ways what this book is about, and is why I'm including Dean's story in this appendix. When I asked Dean what advice he had for filmmakers who want to get their work seen, he responded as follows:

> Make what you want to make, not what you think people want to see. Be prepared to suffer and starve. Don't take the first offer, and get a lawyer to look at anything before you sign it. And plan on making several movies that will not get distro, that will only be seen by your friends, and will just help you hone your art. And don't even look at it as "working toward a payoff." Work like everything you're doing is only for the sake of doing it, because there's a good chance it will be. Enjoy the ride, then if you "get somewhere," it'll be even that much sweeter.

For more on Michael W. Dean, and his most recent project *Hubert Selby, Jr.: It'll Be Better Tomorrow* (a documentary on the author of the books *Requiem for a Dream* and *Last Exit to Brooklyn, www.CubbyMovie.com*), take a look at his Web site: *www.kittyfeet.com*. Dean is also the author of *$30 Film School*, a book on no-budget filmmaking. (*www.30dollarfilmschool.com*)

Don't Sell Yourself Short

When the time comes, remember that you don't have to accept the first contract a distributor gives you. As Dean says above, don't take the first offer—you have every right to negotiate and rework the terms to find a deal that benefits both you and the distributor.

I once saw an interview with Ice Cube, the rapper turned actor, who said the job of a producer is to get him to act in a film for as little money as possible. His job, he said, is to get the producer to pay him as much as possible to act in the same film. Negotiating a distribution contract works pretty much the same way.

It's a common practice for a distribution company, or anyone else you're doing business with, to initially make a lower offer because it then leaves them room to negotiate up. Often, people expect you to make a counter offer and ask for more money upfront as well as a larger percentage of your film's future proceeds. If you don't know that, and accept the first offer someone makes, you miss out on the chance to negotiate a better deal for yourself and may never know what you really could have gotten.

Negotiations are such an important part of any creative endeavor that each semester I teach, I set aside an entire class meeting to work with students on developing a budget and negotiating payment for their work. I start out by asking the class if anyone ever named a dollar figure, either as a buyer or a seller, only to have the person they were negotiating with agree too quickly. Students around the room smile or frown as they relive past experience—we've all been there, you name a price that's too low and the other person just snaps it up.

"Some people become complete pushovers once the film is completed," Micha Green of Cinetic Media told *International Documentary* magazine, the monthly publication of the International Documentary Association. (Cinetic represents individual filmmakers seeking funding and distribution, and also represents large media conglomerates.) "As soon as they're contacted by a company that would like to release their film, suddenly they're all weak in the knees and they basically bend over backwards to accommodate them, forgetting for a brief moment that distributors contact hundreds and hundreds of filmmakers every year."

One suggestion I make to students is to let the other party name a dollar figure first, because if you initiate the bidding you'll never know what they

NOTE

The AIVF Guide to Film and Video Distributors, available online at http://www.aivf.org/ store/, lists contact information and a brief description of "close to 200 film and video distributors," which "is supplemented by company background and advice for indies." AIVF also sells a guide to international film and video festivals, and a self-distribution toolkit, which includes case studies and "successful self-distribution models."

were prepared to offer. The same technique is very valuable in salary negotiations when you start a new job. If you go for an interview and a person asks what your salary requirements are, I suggest asking them what they have budgeted for the position—they may be ready to pay more than you would have asked for.

Of course, your negotiation skills become truly valuable only when people see your work and express an interest in buying it. The first, and most important, step is making a good film. Next, send it to festivals and go to the screenings, schmooze, network, and cajole yourself into situations where people will remember you and your film. You might also consider calling a few distributors and either sending them a screening copy or, better yet, inviting them to a festival that's showing your work—if a distributor sees an audience respond well to your film, it shapes her opinion in a highly positive way. (If you're an established filmmaker with a strong track record, you can also invite potential distributors to a private screening. If you've just finished your first film, this might not be the best way to go, but it could become a good option later in your career once you've gotten some attention.)

At the end of the day, getting your film seen, recognized, and hopefully even purchased takes just as much creativity and effort as making the actual film itself. Once you've cut together something that's truly worth watching, it's time to make sure people see it.

Why Good Business Cards Are Important

When you go to a film festival or a conference, people will often ask you for a business card. A good business card not only enables you to give someone your contact information (remember, you always want to make it easy for people to give you money—and before they can give you money, they need to know where to find you) but it can also make you look professional. If you have a sharp-looking business card, it shows that you pay attention to detail and you're ready to work with other people. On the other hand, if someone asks you for a business card and you come up with a good story about why you don't have any, or if you offer to write your information down on a napkin, it doesn't produce quite the same effect.

A good business card should have your name, email address, and a phone number. (Preferably, this should be a number that won't be answered by an incoherent/irritable roommate or some type of bizarre outgoing message on your voicemail. The goal here is to look professional.) Most of all, a good business card should reflect the image you want to present to the public. You don't need to spend a lot of money on your cards, but everything you produce as a filmmaker should reflect your visual aesthetic and professional competence. Business cards are no exception.

I once worked at a company where the owner had very expensive cards printed up for everyone on the staff, using white ink on a white background. Before the design went to the printer, his wife asked him, in front of the staff, if white ink on a white background would make the text invisible. The boss got very mad, and explained to her that he was the boss and knew what he was doing. Of course, when the cards came back, she was right.

My cards are especially clean and simple (Figure B-1). They contain my name and contact information in a stylish-yet-clear font, and are printed in dark red type on a heavy, natural beige card stock. The type is surrounded by ample areas of blank space, which makes the print-

Ian David Aronson

digitaldocumentary.org
dv@digitaldocumentary.org

Figure B-1. I've spent years trying to come up with the right business card—this is what my current design looks like. (My actual card contains a phone number, which I omitted from this version, but the email address is real.)

ing easier to read and leaves room for people who want to write notes. (In the spring of 2004, I was accepted into a very selective Ph.D program in Educational Communication and Technology at NYU. I had a very sharp set of business cards printed to celebrate, and people frequently compliment the simplicity and elegance of their design.)

If you're not entirely happy with your current set of business cards, have another set printed. As a filmmaker, you never want to put anything out into the world if it doesn't look exactly the way you want it to. That includes your card.

Once you get a good set of business cards that you're happy with, make sure to bring them with you. When I got into my first big film festival back in 1997, I had a box of 250 business cards printed for the occasion and spent hours designing the layout and refining the text. When I got them back from the printer, I felt like a big shot and couldn't wait to hand them out. I then got on a plane to the festival and forgot to bring the cards with me. I still have a thick stack. They provide a very nice souvenir of my early days as a media producer, but they would have done me a lot more good if I had given them to people.

As I was revising a draft of this appendix, and looking for the perfect way to end this book, I walked away from my desk to present *Mi Querida America* at the 2005 NYU Symposium on Technology and Learning. Shortly before my presentation, I connected my laptop to the room's audio video system, and noticed that no sound was coming from the speakers. Since dialog is an important part of my work, this was a cause for concern. As the symposium's staff anxiously ran around connecting and disconnecting cables, pushing buttons, and offering advice, I remembered Jon Else's advice from my film school days and I walked over to the nearest power outlet. As you may have guessed, the speakers were not plugged in.

Else was right, and the advice at the start of this appendix applies to just about any project you'll create.

Make sure your equipment is plugged in. Make a great film, and when you're ready, quit working on it and start showing it to people.

Your life as a filmmaker awaits. Let nothing stop you.

Index

About the Author

Ian David Aronson is a media producer and scholar who lives in New York City. He is a 1997 graduate of the Stanford University Master's Program in Documentary Film and Video, and he is also the director/producer of *digitaldocumentary.org*, which produces social issue documentary and educational media in a variety of electronic formats. His current production, *Mi Querida America*, follows a group of immigrant teens through their first year of high school in Manhattan and is intended for distribution via broadband Internet. From 2001 to 2004, Aronson served as Assistant Professor of Digital Media at Ramapo College. He is currently earning a doctorate in the highly selective Educational Communication and Technology program at New York University. Before attending Stanford he worked as a public radio producer and as a stringer for *The New York Times*.

Colophon

Our look is the result of reader comments, our own experimentation, and feedback from distribution channels. Distinctive covers complement our distinctive approach to technical topics, breathing personality and life into potentially dry subjects.

Darren Kelly was the production editor for *DV Filmmaking: From Start to Finish*. Linley Dolby was the copyeditor. Philip Dangler was the proofreader. Anne Kilgore did the typesetting and page makeup. Genevieve d'Entremont provided quality control. Julie Hawks wrote the index.

Mike Kohnke designed the cover of this book using Adobe Photoshop CS and Adobe InDesign CS, and produced the cover layout with InDesign CS using Linotype Birka and Adobe Myriad Condensed fonts.

David Futato created the series design using Adobe InDesign CS. This book was converted from Microsoft Word to InDesign by Joe Wizda. The text and heading fonts are Linotype Birka and Adobe Myriad Condensed; the sidebar font is Adobe Syntax; and the code font is TheSans Mono Condensed from LucasFont. The illustrations and screenshots that appear in the book were produced by Robert Romano, Jessamyn Read, and Lesley Borash, using Macromedia FreeHand MX and Adobe Photoshop CS.

Better than e-books

Buy *DV Filmmaking: From Start to Finish* and access
the digital edition FREE on Safari for 45 days.

Go to www.oreilly.com/go/safarienabled
and type in coupon code KDGV-ERLF-LSCE-GIF9-PIKC

Search
thousands of
top tech books

Download
whole chapters

Cut and Paste
code examples

Find
answers fast

Search Safari! The premier electronic reference
library for programmers and IT professionals.

39.95

Keep in touch with O'Reilly

Download examples from our books

To find example files from a book, go to: *www.oreilly.com/catalog* select the book, and follow the "Examples" link.

Register your O'Reilly books

Register your book at *register.oreilly.com* Why register your books? Once you've registered your O'Reilly books you can:

- Win O'Reilly books, T-shirts or discount coupons in our monthly drawing.
- Get special offers available only to registered O'Reilly customers.
- Get catalogs announcing new books (US and UK only).
- Get email notification of new editions of the O'Reilly books you own.

Join our email lists

Sign up to get topic-specific email announcements of new books and conferences, special offers, and O'Reilly Network technology newsletters at:

elists.oreilly.com

It's easy to customize your free elists subscription so you'll get exactly the O'Reilly news you want.

Get the latest news, tips, and tools

www.oreilly.com

- "Top 100 Sites on the Web"—PC Magazine
- CIO Magazine's Web Business 50 Awards

Our web site contains a library of comprehensive product information (including book excerpts and tables of contents), downloadable software, background articles, interviews with technology leaders, links to relevant sites, book cover art, and more.

Work for O'Reilly

Check out our web site for current employment opportunities:

jobs.oreilly.com

Contact us

O'Reilly Media, Inc.
1005 Gravenstein Hwy North
Sebastopol, CA 95472 USA
Tel: 707-827-7000 or 800-998-9938
      (6am to 5pm PST)
Fax: 707-829-0104

Contact us by email

For answers to problems regarding your order or our products:
order@oreilly.com

To request a copy of our latest catalog:
catalog@oreilly.com

For book content technical questions or corrections: **booktech@oreilly.com**

For educational, library, government, and corporate sales: **corporate@oreilly.com**

To submit new book proposals to our editors and product managers:
proposals@oreilly.com

For information about our international distributors or translation queries:
international@oreilly.com

For information about academic use of O'Reilly books:
adoption@oreilly.com
or visit:
academic.oreilly.com

For a list of our distributors outside of North America check out:
international.oreilly.com/distributors.html

Order a book online

www.oreilly.com/order_new

**O'REILLY**®   Our books are available at most retail and online bookstores.
To order direct: 1-800-998-9938 • *order@oreilly.com* • *www.oreilly.com*
Online editions of most O'Reilly titles are available by subscription at *safari.oreilly.com*